Biotechnology Information Science and Regulatory System

Biotechnology Information Science and Regulatory System

P. R. Gilbert

ANMOL PUBLICATIONS PVT. LTD.
NEW DELHI - 110 002 (INDIA)

ANMOL PUBLICATIONS PVT. LTD.
H.O.: 4374/4B, Ansari Road, Darya Ganj
New Delhi-110 002 (India)
Ph.: 23278000, 23261597
B.O.: 1015, Ist Main Road, BSK IIIrd Stage
IIIrd Phase, IIIrd Block
Bangalore - 560 085 (Karnataka)
Ph: 080-41723429, Tel/Fax: 080-2672 3604
Visit us at: www.anmolpublications.com

Biotechnology Information Science and Regulatory ***System***

First Published, 2008

ISBN 978-81-261-3282-9

PRINTED IN INDIA

Printed at Mehra Offset Press, Delhi.

Contents

Preface

This book entitled "Biotechnology Information Science and Regulatory System" is designed to acquaint readers with the biotechnology information science and regulatory system at the global level. Bioinformatics is an interdisciplinary field that addresses biological problems using computational techniques. The field is also often referred to as computational biology. It plays a key role in various areas like functional genomics, structural genomics, and proteomics amongst others, and forms a key component in biotechnology and pharmaceutical sector. Bioinformatics and computational biology involve the use of techniques from applied mathematics, informatics, statistics, and computer science to solve biological problems. Research in computational biology often overlaps with systems biology. Major research efforts in the field include sequence alignment, gene finding, genome assembly, protein structure alignment, protein structure prediction, prediction of gene expression and protein-protein interactions, and the modeling of evolution. The terms *bioinformatics* and *computational biology* are often used interchangeably, although the former typically focuses on algorithm development and specific computational methods, while the latter focuses more on hypothesis testing and discovery in the biological domain. Although this distinction is used by National Institutes of Health in their working definitions of Bioinformatics and Computational Biology, it is clear that there is a tight coupling of developments and knowledge between the more hypothesis-driven research in computational biology and technique-driven research in bioinformatics. Computational biology also includes lesser known but equally important sub-disciplines such as computational biochemistry and computational biophysics. A common thread in projects in bioinformatics and computational biology is the use of mathematical tools to extract useful information from noisy data produced by high-throughput biological techniques such as genomics (The field of data mining overlaps with computational biology in this regard).

A representative problem in bioinformatics is the assembly of high-quality DNA sequences from fragmentary "shotgun" DNA sequencing, while in computational biology, a representative problem might be statistical testing of a hypothesis of common gene regulation using data from mRNA microarrays or mass spectrometry. Since the Phage Ö-X174 was sequenced in 1977, the DNA sequences of more and more organisms have been decoded and stored in electronic databases. This data is analyzed to determine genes that code for proteins, as well as regulatory sequences. A comparison of genes within a species or between different species can show similarities between protein functions, or relations between species (the use of molecular systematics to construct phylogenetic trees). With the growing amount of data, it long ago became impractical to analyze DNA sequences manually. Today, computer programs are used to search the genome of thousands of organisms, containing billions of nucleotides. These programs can compensate for mutations (exchanged, deleted or inserted bases) in the DNA sequence, in order to identify sequences that are related, but not identical. A variant of this sequence alignment is used in the sequencing process itself. The so-called shotgun sequencing technique (which was used, for example, by The Institute for Genomic Research to sequence the first bacterial genome, *Haemophilus influenza*) does not give a sequential list of nucleotides, but instead the sequences of thousands of small DNA fragments (each about 600-800 nucleotides long). The ends of these fragments overlap and, when aligned in the right way, make up the complete genome. Shotgun sequencing yields sequence data quickly, but the task of assembling the fragments can be quite complicated for larger genomes.

On the other hand, the biotechnology regulatory system includes things such as the marketing of GM food is regulated in a manner that helps to provide the necessary levels of safety, transparency and reassurance. At the beginning of the 2000's, officials insisted that new regulations were needed to "restore consumer confidence" in the technology. These new regulations required strict labelling and traceability of all food and animal feed containing more than 0.5 % GM ingredients. Directives, such as directive 2001/18/EC, were designed to require authorisation for the placing on the market of GMO, in accordance with the precautionary principle. One of the features of this system is a comprehensive pre-market risk assessment, a system trying to provide means for products to be followed at each stage of their production and distribution, by both transmission of accurate information and labelling. This traceability is a means to implement post-market measures such as monitoring and withdrawals (recalls).

This system is not only limited to GMO products but should encompass any food product ultimately. For example, the original EU rules for labelling of GM products were limited to products where transformed DNA and/or transformed protein are detectable, not to products that have been produced from GMOs but no longer appears to contain modified DNA and/or proteins. New rules for tracebility and labelling which came into force in 2004 also require labelling of highly refined products made from GM indgredirents like oil and corn syrup, even though that the presence of recombinant DNA or protein cannot be proven. The labelling rules do not apply to products of microbial genetic engineering, so the cheese made with the help of GM-chymosin doesn't have to be labelled. Officials stress that while traceability facilitates the implementation of safety measures, where appropriate, it cannot and should not be considered as a safety measure. In 1999, a 4 year ban was pronounced on new genetically modified crops. At the end of 2002, European Union environment ministers agreed new controls on GMOs could eventually lead the 25-member bloc to reopen its markets to GM foods. European Union ministers agreed to new labelling controls for genetically modified goods which will have to carry a special harmless DNA sequence (a DNA code bar) identifying the origin of the crops, making it easier for regulators to spot contaminated crops, feed, or food, and enabling products to be withdrawn from the food chain should problems arise. A series of additional sequences of DNA with encrypted information about the company or what was done to the product could also be added to provide more data.

This book consists of—Biotech Information and Database: An Introductory Overview; Institutions Related to Biotechnology Information; Biotech Information Mapping; Biotechnology Research and Regulations; Biotechnology and Telecommunication: Modern Innovations; and Code of Conduct of Biotechnology etc. besides a large body of list of acronyms, glossary of relevant terms, extensive bibliography, list of websites and links, for further referencing and research.

—Editor

List of Acronyms

ACS	—	American Chemical Society
ADA	—	Americans with Disabilities Act
AEC	—	Atomic Energy Commission
AES	—	American Electrophoresis Society
AFMR	—	American Federation of Medical Research
AFIP/ARP	—	Armed Forces Inst. of Pathol./Am. Registry of Pathol.
AGSG	—	Alliance of Genetic Support Groups (now Genetic Alliance)
AGT	—	Association of Genetic Technologists
AHA	—	American Heart Association
AIBS	—	Am. Inst. of Biol. Soc.
AMIA	—	American Medical Informatics Association
BDG	—	Batten's disease gene
BER	—	Biological and Environmental Research
BIO	—	Biotechnology Industry Organization
BNL	—	Brookhaven National Laboratory
BSCS	—	Biological Sciences Curriculum Study
BS/SCF	—	Biological Sequence/Structure Computational Facility

BTCI	—	BioPharmaceutical Technology Center Institute
BTP	—	Biotechnology Training Programs
CAE	—	capillary array electrophoresis
CaSSS	—	California Separation Science Society
CATCMB	—	Center for Advanced Training in Cell and Molecular Biology
CLMA	—	Clinical Laboratory Management Association
CMT	—	Charcot-Marie-Tooth
CONTIG	—	Consortium of Teachers in Genetics
CORN	—	Council of Regional Networks for Genetic Services
CRADA	—	Cooperative Research and Development Agreement
CSHL	—	Cold Spring Harbor Laboratory
CTG	—	Trinucleotide Repeat
DHHS	—	Dept. of Health and Human Services
DIMACS	—	Center for Discrete Math. & Theoretical Comp. Sci.
DM	—	Myotonic Dystrophy
DMD	—	Duchenne Muscular Dystrophy
DNA	—	Deoxyribonucleic acid
DOE	—	Department of Energy
ECB	—	European Congress on Biotechnology
EDS	—	Electronic Data Submission
EEOC	—	Equal Employment Opportunity Commission
EL.B.A.	—	ELectronics and Biotechnology Advanced

EMBL	—	European Mol. Biol. Lab.
EMBO	—	European Molecular Biology Organisation
EMG	—	Encyclopedia of the Mouse Genome
EMS	—	Environmental Mutagen Society
EORTC	—	European Organization for Research and Treatment of Cancer
ERDA	—	Energy Research and Development Administration
ERI	—	Eleanor Roosevelt Institute
ES	—	Embryonic Stem
ESF	—	European Science Foundation
ESHG	—	European Society of Human Genetics
ESI	—	Electrospray Ionization
EST	—	Expressed Sequence Tag
EUCIB	—	European Collaborative Interspecific Backcross
EURESCO	—	European Research Conferences
FAP	—	Familial Adenomatous Polyposis
FASEB	—	Federation of American Societies for Experimental Biology
FCM	—	Flow Cytometry
FEBS	—	Fed. of Eur. Biochem. Soc.
FMF	—	Familial Mediterranean Fever
FRAXA	—	Fragile X locus
FSHD	—	Facioscapulohumeral Muscular Dystrophy
FTICR	—	Fourier Transform Ioncyclotron Resonance

FVEA	—	Fundacion Valenciana de Estudios Avanzados
GAS	—	Genome Automation System
GBASE	—	Genome Database of the Mouse
GBR	—	Global Business Research
GDB	—	Genome Database
GDB/OMIM	—	Genome Database/Online Mendelian Inheritance in Man
GESTEC	—	Genome Science and Technology Center
GIST	—	Genome Informatics System of Transputers
GLaRGG	—	Great Lakes Regional Genetics Group
GMCRF	—	General Motors Cancer Res. Foundation
GMD	—	Genomic Map Design
GPI	—	Genetics and Public Issues Program
GRAIL	—	Gene Recognition and Analysis Internet Link
GRC	—	Gordon Res. Conf.
GSA	—	Genetics Societies of America
GSDB	—	Genome Sequence Data Base
HAEC	—	Human Artificial Episomal Chromosome
HELSRD	—	Health Effects and Life Science Research Division
HERAC	—	Health and Environmental Research Advisory Committee
HGCC	—	Human Genome Coordinating Committee

HGM	—	Human Genome Meeting
HGMIS	—	Human Genome Management Information System
HGP	—	Human Genome Project
HHMI	—	Howard Hughes Medical Institute
HICSS	—	Hawaii Intl. Conf. on Systems Sci.
HLA	—	Human Leukocyte Antigen
HMDP	—	Homology Database
HNPCC	—	Hereditary Nonpolyposis Colorectal Cancer
HSA	—	Human Serum Albumin
HUGO	—	Human Genome Organisation
IARC	—	International Agency for Research on Cancer
IBC	—	International Business Communications
IBEX	—	International Biotechnology EXpo
IBF	—	International Business Forum
IBI	—	Institute for Biotechnology Information
ICES	—	International Council of Electrophoresis Societies
ICGEB	—	International Centre for Genetic Engineering and Biotechnology
ICHG	—	International Congress of Human Genetics
ICPEMC	—	International Commission on Protection Against Environmental Mutagens and Carcinogens
ICRF	—	Imperial Cancer Research Fund

IEEE	—	Institute of Electrical and Electronics Engineers
IG	—	IntelliGenetics
IGES	—	International Genetics Epidemiology Societies
IJCAI	—	International Joint Conference on Artificial Intelligence
IMA	—	Institute for Mathematics and its Applications
IMACS	—	International Association for Mathematics and Computers In Simulation
IMAGE	—	Integrated Molecular Analysis of Gene Expression
IMGS	—	Intl. Mammalian Genome Society
INRIA	—	French Natl. Inst. for Research in Computer Science and Control
IOM	—	Institute of Medicine
IOR	—	Institute of Religion
ISAG	—	Intl. Society for Animal Genetics
ISMB	—	Intelligent Systems for Molecular Biology
ISONG	—	Intl. Soc. of Nurses in Genet.
ISQL	—	Interactive Standard Query Language
ISTR	—	Institute for Science Training and Research
IUBMB	—	Intl. Union of Biochemistry and Molecular Biology
IU	—	Indiana University
IVD	—	*in vitro* Diagnostic
JGI	—	Joint Genome Institute

LANL	—	Los Alamos National Laboratory
LBNL	—	Lawrence Berkeley National Laboratory
LLNL	—	Lawrence Livermore National Laboratory
LTI	—	Life Technologies, Inc.
MALDI	—	matrix-assisted laser desorption ionization
MBC	—	Massachusetts Biotechnology Council
MBL	—	Marine Biological Laboratory
MCD	—	Mouse Cytogenetic Database
MDA	—	Muscular Dystrophy Assoc.
MEMS	—	Microelectromechanical Systems
MGC	—	Mouse Genome Conference
MGD	—	Mouse Genome Database
MGI	—	Microbial Genome Initiative
MHC	—	Major Histocompatibility Complex
MIMBD	—	Meet. on the Interconnection of Mol. Biol. Databases
MIT	—	Massachusetts Institute of Technology
MMRF	—	Marshfield Medical Research Foundation
MNBWS	—	Miami Nature Biotechnology Winter Symposium
MOD	—	March of Dimes
MOU	—	Memorandum of Understanding
MRC	—	Medical Research Council
MS	—	Mass Spectrometry
NACHGR	—	Natl. Advisory Council for Human

		Genome Res.
NAPBC	—	Natl. Action Plan on Breast Cancer
NAS	—	Natl. Academy of Sciences
NASA	—	Natl. Aeronautics and Space Administration
NBAC	—	Natl. Bioethics Advisory Commission
NCGR	—	National Center for Genome Resources
NCI	—	Natl. Cancer Institute
NCSA	—	National Center for Supercomputing Applications
NCSL	—	National Conference of State Legislatures
NCSU	—	North Carolina State University
NEH	—	National Endowment for the Humanities
NFCR	—	National Foundation for Cancer Research
NFID	—	National Foundation for Infectious Diseases
NHGRI	—	Natl. Human Genome Res. Inst.
NICHD	—	National Institute of Child Health and Human Development
NIGMS	—	National Institute of General Medical Sciences
NIH	—	National Institutes of Health
NIGMS	—	Natl. Institute of General Medical Sciences
NIST	—	Natl. Inst. of Standards and Technology

NLGLP	—	National Laboratory Gene Library Project
NLM	—	National Library of Medicine
NMHCC	—	National Managed Health Care Congress
NORD	—	National Organization for Rare Disorders
NRC	—	National Research Council
NRSA	—	National Research Service Award
NSF	—	Natl. Sci. Foundation
NSGC	—	National Society of Genetic Counselors
NSTA	—	National Science Teachers Association
NYAS	—	New York Academy of Science
OAINN	—	Ohio Aerospace Institute Neural Networks
OBER	—	Office of Biological and Environmental Research
OHER	—	Office of Health and Environmental Research
OMIM	—	Online *Mendelian Inheritance in Man*
OPRR	—	Office of Protection from Res. Risks
ORAU	—	Oak Ridge Associated Universities
ORISE	—	Oak Ridge Inst. for Science and Education
ORF	—	open reading frame
ORNL	—	Oak Ridge National Laboratory
OSU	—	Oregon State University
OTA	—	Office of Technology Assessment

PACHG	—	Program Advisory Committee on the Human Genome
PCR	—	Polymerase Chain Reaction
PFGE	—	Pulsed-field gel electrophresis
PG	—	Plant Genome
RARA	—	Retinoic Acid Receptor
RECOMB	—	Conference on Computational Molecular Biology
RFLP	—	Restriction Fragment Length Polymorphism
RH	—	Radiation Hybrid
RLGS	—	Restriction Landmark Genomic Scanning
SASE	—	Sample Sequencing
SBH	—	Sequencing by Hybridization
SBIR	—	Small Business Innovation Research
SCAN	—	Sequence Comparison ANalysis Program
SCE	—	School of Continuing Education
SCI	—	Society of Chemical Industry
TIGR/NIST	—	The Inst. for Genomic Res./Natl. Inst. of Standards and Technol.
UC	—	University of California
UCB	—	University of California, Berkeley
WCCS	—	Workshop on Complete cDNA Sequencing

List of Glossary of Terms

Amino acid: Any of 20 basic building blocks of proteins—composed of a free amino (NH2) end, a free carboxyl (COOH) end, and a side group (R).

Ampicillin (beta-lactamase): An antibiotic derived from penicillin that prevents bacterial growth by interfering with cell wall synthesis.

Amplify: To increase the number of copies of a DNA sequence, in vivo by inserting into a cloning vector that replicates within a host cell, or in vitro by polymerase chain reaction (PCR).

Anaerobe: An organism that grows in the absence of oxygen. See Aerobe.

Analyte: A chemical species targeted for qualitative or quantitative analysis.

Anneal: The pairing of complementary DNA or RNA sequences, via hydrogen bonding, to form a double-stranded polynucleotide. Most often used to describe the binding of a short primer or probe.

Antibiotic resistance: The ability of a microorganism to produce a protein that disables an antibiotic or prevents transport of the antibiotic into the cell.

Antibiotic: A class of natural and synthetic compounds that inhibit the growth of or kill other microorganisms.

Antibody: An immunoglobulin protein produced by B-lymphocytes of the immune system that binds to a specific antigen molecule.

Anticodon: A nucleotide base triplet in a transfer RNA molecule that pairs with a complementary base triplet, or codon, in a messenger RNA molecule. See Codon, Messenger RNA, RNA.

Antigen: Any foreign substance, such as a virus, bacterium, or protein, that elicits an immune response by stimulating the production of antibodies.

Bacteriophage (phage or phage particle): A virus that in- fects bacteria. Altered forms are used as vectors for cloning DNA.

Bacteriophage (phage): Any virus that infects bacteria. They were the first organisms used for the study of molecular genetics and are now widely used as cloning vectors. (From Greek *phagein*, to eat.)

Bacteriorhodopsin: Pigmented protein found in the plasma membrane of a salt-loving bacterium, *Halobacterium halobium*; it pumps protons out of the cell in response to light.

Bacteriostat: A class of antibiotics that prevents growth of bacterial cells.

Bacterium (plural bacteria): Common name for any member of the diverse group of procaryotic organisms. Most are single cells, but multicellular forms also exist.

Bacterium: A single-celled, microscopic prokaryotic organism: a single cell organism without a distinct nucleus.

Band: In gels and blots, bands are visible indications of a particular fragment of a certain size. There may be one to many bands per lane.

Banoprism: Scientists at Northwestern University have created a nanoparticle with a new shape that could be a useful tool in the race to detect biological threats. The nanoprism, which resembles a tiny Dorito, exhibits unusual optical properties that could be used to improve biodetectors, allowing them to test for a far greater number of biological warfare agents or diseases at one time.

Banoscience: The study of phenomena and manipulation of materials at atomic, molecular and macromolecular scales, where properties differ significantly from those at a larger scale. Nanoscience is primarily the extension of existing sciences into the realms of the extremely small (*nanomaterials, nanochemistry, nanobio, nanophysics*, etc.) while *nanoengineering* represents the extension of the engineering fields into the nano- scale realm (*nanofabrication, nanodevices*, etc.). The exponential growth of nanoscience is largely due to the development of new instruments and related techniques that are used to "routinely" probe and manipulate material at the atomic and molecular level. *Scanning probe microscopies,* analytical electron- beam techniques, epitaxial growth facilities, and *synchrotron* radiation sources are all opening huge opportunities.

Banoshells: Procedures that target cancer cells while leaving normal cells untouched, patients controlling the release of medicine in their bodies with an infrared light, and medical test results produced in seconds rather than days – these are three new medical technologies currently being tested by nanotechnology researchers at Rice University. ... The research focuses on nanoshells, a new type of nanoparticle invented by Naomi Halas, Rice professor of electrical and computer engineering and chemistry. Nanoshells are layered nanoparticles whose ability to manipulate light and color can be designed into the nanoparticle by varying the thickness of the nanoparticle's layers.

Basal body: Short cylindrical array of microtubules plus their associated proteins found at the base of a eucaryotic cell cilium or flagellum. Serves as a nucleation site for the growth of the axoneme. Closely similar in structure to a centriole.

Basal lamina (plural basal laminae): Thin mat of extracellular matrix that separates epithelial sheets, and many types of cells such as muscle cells or fat cells, from connective tissue. Sometimes called a basement membrane.

Base pair (bp): A pair of complementary nitrogenous bases in a DNA molecule—adenine-thymine and guanine-cytosine. Also, the unit of measurement for DNA sequences. In DNA, there are four possible bases: cytosine (C), guanine (G), adenine (A), and thymine (T). Cytosine and thymine are pyrimidine bases; adenine and guanine are purine bases. Cytosine is complementary to guanine while adenine is complementary to thymine. If one strand of DNA has the sequence ATTGC then the complementary strand will be TAACG. Two complementary bases constitute a base pair. In RNA, thymine is replaced by uracil.

Beta-DNA: The normal form of DNA found in biological systems, which exists as a right-handed helix.

Beta-Lactamase: Ampicillin resistance gene.

Binding energy: The reduction in the free energy of a system that occurs when a ligand binds to a receptor. Generally used to describe the total energy required to remove something, or to take a system apart into its constituent particles—for example, to separate two atoms from one another, or to separate an atom into electrons and nuclei.

Bio-assemblies or Biomolecular Assemblies: Containing

several protein units, DNA loops, lipids, various ligands, etc.

Bioaugmentation: Increasing the activity of bacteria that decompose pollutants; a technique used in bioremediation.

Biochauvinism: The prejudice that biological systems have an intrinsic superiority that will always give them a monopoly on self-reproduction and intelligence.

Biocompatible coated materials: Biocompatible materials usually used in dental and bone implants that enhance biologic fixation, thereby increasing the bond strength between the coated material and bone, and minimize possible biological effects that may result from the implant itself. MeSH, 1999

Biocompatible materials: Synthetic or natural materials, other than drugs, that are used to replace or repair any body tissue or bodily function. MeSH, 1973

Biodefense: The need for improved national defenses against biological attacks are urgently needed. A real-time reporting system needs to be developed, whereby officials can be informed about an emerging threat before an agent has a chance to affect thousands of people. The development of integrated systems for detecting and monitoring biological agents is instrumental to this goal. The development of vaccines and novel detection platforms will enable preparedness, while also benefiting human health.

Biodiversity: The wide diversity and interrelatedness of earth organisms based on genetic and environmental factors.

Biodynotics Biologically Inspired Multifunctional Dynamic Robotics: A multidisciplinary, multi- pronged approach with far reaching impact on robotic capabilities for national security applications. Biologically Inspired Multifunctional Dynamic Robotics: BIODYNOTICS will explore the following areas: DYNAMIC MOBILITY, BEHAVIOR, INTEGRATION Biological Sciences, Defense Sciences Office, DARPA.

Bioengineering: Is rooted in physics, mathematics, chemistry, biology, and the life sciences. It is the application of a systematic, quantitative, and integrative way of thinking about and approaching the solutions of problems important to biology, medical research, clinical proactive, and population studies. The NIH Bioengineering Consortium agreed on the following definition for bioengineering research on biology, medicine, behavior, or health recognizing that no definition could completely eliminate overlap with other research disciplines or preclude variations in interpretation by different individuals and organizations. Integrates physical, chemical, or mathematical sciences and engineering principles for the study of biology, medicine, behavior, or health. It advances fundamental concepts, creates knowledge for the molecular to the organ systems levels, and develops innovative biologics, materials, processes, implants, devices, and informatics approaches for the prevention, diagnosis, and treatment of disease, for patient rehabilitation, and for improving health.

Bioenrichment: Adding nutrients or oxygen to increase microbial breakdown of pollutants.

Biofabrication: Includes nanoparticle delivery systems, biomaterials, tissue engineering, implants and prostheses. Using biological processes to synthesize and manufacture chemicals and materials of high value to the Department of Defense. Biological processes are characterized by: low energy barriers (~10Kcal, high temperatures/pressures not required); high reaction-, regio- and stereo- specificity (protection/ deprotection wastes and catalyst poisoning avoided); spatio-temporal control of materials synthesis of defined composition and size with angstrom- level precision; and local control of the dielectric environment (largely eliminating the need for toxic solvents). Potential target materials and chemicals include composites with enhanced mechanical properties, ultra-low k dielectrics and thermoelectrics, optoelectronic materials, photonic devices (waveguides and logic elements), electronic materials (e.g., GaN, InGaN, AlGaN), elastomers, energetic materials, and adhesives. Biofabrication, DARPA.

Biologics: Agents, such as vaccines, that give immunity to diseases or harmful biotic stresses.

Biomass: The total dry weight of all organisms in a particular sample, population, or area.

Biomaterials: Synthetic or natural materials that can replace or augment tissues, organs or body functions.

Biomechanics: Mechanical structures of living organisms (especially muscles and bones).

Biomedical Nanotechnology: (1) the comprehensive monitoring, control, construction, repair, defense, and improvement of all human biological systems, working from the molecular level, using engineered nanodevices and nanostructures; (2) the science and technology of diagnosing, treating, and preventing disease and traumatic injury, of relieving pain, and of preserving and improving human health, using molecular tools and molecular knowledge of the human body; (3) the employment of molecular machine systems to address medical problems, using molecular knowledge to maintain and improve human health at the molecular scale. (4) A rapidly expanding field that includes many potential technologies and approaches. The key to this definition is that phenomena and materials at the nanometer scale are known to have properties that are uniquely attributable to that scale length. Nanomedicine could similarly be defined [as nanotechnology] as the design, synthesis, or application of materials, devices, or technologies in the nanometer-scale for the basic understanding, diagnosis, and / or treatment of disease. Canadian Institute of Health Research, Regenerative Medicine and Nanomedicine, RFA, 2003. The monitoring, repair, construction and control of human biological systems at the molecular level, using engineered nanodevices and nanostructures. (5) the comprehensive monitoring, control, construction, repair, defense, and improvement of all human biological systems, working from the molecular level, using engineered nanodevices and nanostructures; (6) the science and technology of diagnosing, treating, and preventing disease and traumatic injury, of relieving pain, and of preserving and improving human health, using molecular tools and molecular knowledge of the human body; (7) the employment of molecular machine systems to address medical problems, using molecular knowledge to maintain and improve human health at the molecular scale. The use of molecular-scale devices (nanotechnology) to repair damage and boost the immune system. Vapour grown carbon fibres are obtained as shown in this SEM image. The diameters of these fibres can vary from 100 nm to 500 nm. As part of the National Institutes of Health (NIH) Roadmap for Medical Research [nihroadmap.nih. gov], the NIH [nih.gov] will establish a handful of nanomedicine centers. These centers will be staffed by a highly interdisciplinary scientific crew including biologists, physicians, mathematicians, engineers and computer scientists. Research conducted over the first few years will be spent gathering extensive information about how molecular machines are built. A key activity during this time will be the development of a new kind of vocabulary, or lexicon, to define biological parts and processes in engineering terms. Once researchers have completely catalogued the interactions between and within molecules, they can begin to look for patterns and a higher order of connectedness than is possible to identify with current experimental methods. Mapping these networks and understanding how they change over time will be a crucial step toward helping scientists understand nature's rules of biological design. Understanding these rules will, in many years' time, enable researchers to use this information to address biological issues in unhealthy cells.

Biomedical polymers: More and more therapeutic problems are relevant to the use of polymer- based therapeutic aids for a limited period of time, namely the healing time related to the outstanding capacity of living systems to self- repair ... After healing the remaining prosthetic materials or devices become foreign residues or wastes that have to be eliminated from the body. Nowadays, biocompatible polymers that can degrade in the body are developed. The degradation and the elimination of degradation by-products depend on rather complex phenomena that are presently reflected inconsistently by terms issued from the tradition because each domain has developed its own terminology almost independently. This is a source of misunderstandings, confusions and misperceptions among scientists, surgeons, pharmacists, journalists and politicians, the situation being increased by the introduction of degradable polymers in plastic waste management and environmental protection.

Therefore, it is urgent to reflect the various phenomena by specific terms, harmonize and enforce their use by the people active in the biomedical, pharmacological and environmental fields, and, last but not least by the publishing media and journalists. IUPAC, Terminology for biomedical (therapeutic) polymers.

BioMEMS Biological MicroElectro Mechanical Systems: Highlights the technical advances in the field that are leading a revolution in medicine, and creating a new generation of analytical devices for medical diagnostics. The meeting will encompass technology developments in micro & nano drug delivery, interface of nanotech and tissue engineering, microfluidics, and miniaturized total analysis systems (microTAS), biosensors, innovations in mass spec, and nanoscale imaging *BioMEMS and Nanotech World*, Aug. 16- 17, 2004, Washington DC

BioMEMS: MEMS used in medicine, that use microchips.

Biomimetic Chemistry: Knowledge of biochemistry, analytical chemistry, polymer science, and biomimetic chemistry is linked and applied to research in designing new molecules, molecular assemblies, and macromolecules having biomimetic functions. These new bio-related materials of high performance, including, for example, enzyme models, synthetic cell membranes, and biodegradable polymers, are prepared, tested, and constantly improved in this division for industrial scale production.

Biomimetic materials: Materials fabricated by BIOMIMETICS techniques, i.e., based on natural processes found in biological systems. [MeSH 2003] (2) Materials that imitate, copy, or learn from nature.

Biomimetic: Imitating, copying, or learning from nature.

Nanotechnology already exists in nature; thus, nanoscientists have a wide variety of components and tricks already available. An interdisciplinary field in materials science, ENGINEERING, and BIOLOGY, studying the use of biological principles for synthesis or fabrication of BIOMIMETIC MATERIALS. There is a need to develop the next generation of restorative materials and medical implants. New avenues of scientific inquiry may enable the development of biomaterials that are safe, reliable, "smart", long-lasting, and perform ideally in their respective biological environments.... Over the last few years biomimetics and tissue engineering have emerged as a new vision in the field of tissue and organ repair and restoration. Biomimetics and tissue engineering are interdisciplinary fields that combine information from the study of biological structures and their functions with physics, mathematics, chemistry and engineering for the generation of new materials, tissues and organs. In the area of craniofacial, oral and dental principles from biomimetics and tissue engineering are applied to developing dental and facial implants, new polymers for guided tissue regeneration used in treating periodontal disease and bone and connective tissue defects, coral- based hydroxyapatite replicas for reconstruction of alveolar ridges and other osseous defects, temporomandibular joint (TMJ) and other joint prostheses, formation of bone matrix substitutes, and artificial replicas of bone, skin, and mucosa. The term biomimetics was coined in 1972 in the context of artificial enzymes; it might be defined broadly as "the abstraction of good design from nature". It is a fact of everyday life that nature has managed to built materials and 'devices' with breathtaking functionality, heterogeneity and stability by using a comparatively limited number of building blocks (the whole range of synthetic materials is restricted to man-made engineering). The basic concepts of nature are often simple; it is the way in which building blocks and materials are arranged that results in functionality. Among the most simple and abundant themes of nature is self- assembly: lipids assemble in sheets to form cell membranes, proteins assemble into functional enzymes, cellular 'sensors', fibers, or virus coats, and DNA assembles in double strands to provide the very basis for live: replication. Biomimetics Study of the structure and function of biological substances to make artificial products that mimic the natural ones.

BIOML Biopolymer Markup Language: Designed by the BIOML core team at Proteometrics, LLC and Proteometrics Canada Ltd. It is to be used "for the annotation of biopolymer sequence information. BIOML allows the full specification of all experimental information known about molecular entities composed of biopolymers, for example, proteins and genes."

Biomolecular materials: An emerging discipline, materials whose properties are abstracted from biology. They share many of the characteristics of biological materials but are not necessarily of biological origin. For example, they may be inorganic materials that are organized or processed in a biomimetic fashion. A key feature of biological and biomolecular materials is their ability to undergo *self- assembly*.

Biomolecular Nanotechnology: Nanotechnology existing in living systems and resulting from our ability to use biomolecules as components for molecular nanotechnology.

Biomotors: Driven by energy sources such as *adenosine triphosphate (ATP)* for chemical transduction and other processes. These biomotors are considered to be biomolecular and are discussed in the body of this report, but strictly speaking they do not conform to the panel's definition of self- assembly.

Bionanotechnology: Includes molecular motors, biomaterials, single molecule manipulation technologies, biochip technologies, etc.

Biopolymeroptoelectromechanical Systems [BioPOEMS]: Combining optics and microelectromechanical systems, and used in biological applications.

Biopolymers: Macromolecules (including proteins, nucleic acids and polysaccharides) formed by living organisms.

Bioremediation: The use of microorganisms to remedy environmental problems. See Bioaugmentation, Bioenrichment.

Biorobotics: Our research focuses on the role of sensing and mechanical design in motor control, in both robots and humans. This work draws upon diverse disciplines, including biomechanics, systems analysis, and neurophysiology. The main approach is experimental, although analysis and simulation play important parts. In conjunction with industrial partners, we are developing applications of this research in biomedical instrumentation, teleoperated robots, and intelligent sensors.

Biosensor: The term "biosensor" is a general designation that denotes either a sensor to detect a biological substance or a sensor which incorporates the use of biological molecules such as antibodies or enzymes. Biosensors are a subcategory of chemical sensors.

Biostasis: A condition in which an organism's cell and tissue structure are preserved, allowing later restoration by cell repair machines. Applicable to cryonics.

Biotechnology: The scientific manipulation of living organisms, especially at the molecular genetic level, to produce useful products. Gene splicing and use of recombinant DNA (rDNA) are major techniques used. Technology for working with biological systems. Includes genetic engineering, human and veterinary medicine, crop and animal breeding, diagnostics, pharmaceuticals, forensics, etc. Narrow sense: Genetic engineering.

Biotic stress: Living organisms which can harm plants , such as viruses, fungi, and bacteria, and harmful insects. See Abiotic stress.

Biovorous: From "biovore;" an organism capable of converting biological material into energy for sustenance.

Bipolar-junction transistor: Transistor with n-type and p-type semiconductors having base-emitter and collector-base junctions.

Blocks Substitution Matrix: A substitution matrix in which scores for each position are derived from *observations* of the frequencies of substitutions in blocks of local alignments in related proteins. Each matrix is tailored to a particular evolutionary distance. In the BLOSUM62 matrix, for example, the alignment from which scores were derived was created using sequences sharing no more than 62% identity. Sequences more identical than 62% are represented by a single sequence in the alignment so as to avoid over-weighting closely related family members. (Henikoff and Henikoff)

Blot, northern: A pattern of RNA fragments transferred to a nitrocellulose membrane from a gel. The gel has undergone electrophoresis to separate fragments according to size. The RNA fragments are arranged in different lanes for each sample. Each lane contains bands which are fragments of different sizes. Radioactive probes are often used to visualized particular bands by autoradiography.

Blot, southern: A pattern of DNA fragments transferred to a nitrocellulose membrane from a gel. The gel has undergone electrophoresis to separate fragments according to size. The DNA fragments are arranged in different lanes for each sample. Each lane contains bands which are fragments of different sizes. Radioactive probes are often used to visualized particular bands by autoradiography.

Blot, western: A pattern of proteins transferred to a nitrocellulose membrane from a gel. The gel has undergone electrophoresis to separate fragments according to size. The proteins are arranged in different lanes for each sample. Each lane contains bands which are fragments of different sizes. Labeled antibody probes are often used to visualize particular bands.

Bone substitutes: Synthetic or natural materials for the replacement of bones or bone tissue. They include hard tissue replacement polymers, natural coral, hydroxyapatite, beta- tricalcium phosphate, and various other biomaterials. The bone substitutes as inert materials can be incorporated into surrounding tissue or gradually replaced by original tissue.

Brownian Assembly: Brownian motion in a fluid brings molecules together in various position and orientations. If molecules have suitable complementary surfaces, they can bind, assembling to form a specific structure. Brownian assembly is a less paradoxical name for self-assembly (how can a structure assemble itself, or do anything, when it does not yet exist?).

Bt (*Bacillus thuringiensis*): A soil bacterium that produces insecticidal proteins. There are several different kinds of proteins produced by different strains of Bt. Some are effective against larvae of moths and butterflies. Others are effective against larvae of beetles. The Bt protein has been introduced into various crops as a built-in insecticide.

Calorie: Unit of heat. One calorie (small "c") is the amount of heat needed to raise the temperature of 1 gram of water by 1 °C. A kilocalorie is the unit used to describe the energy content of foods.

CAP (catabolite gene activator protein): Gene regulatory protein in procaryotes that, when glucose is absent, activates genes responsible for the breakdown of alternative carbon sources.

Capacitor: Energy storage circuit element having two conductors separated by an insulator.

Carcinogen: A substance that induces cancer.

Carcinoma: A malignant tumor derived from epithelial tissue, which forms the skin and outer cell layers of internal organs.

Catalyst: A substance that promotes a chemical reaction by lowering the activation energy of a chemical reaction, but which itself remains unaltered at the end of the reaction.

Catalytic antibody (abzyme): An antibody selected for its ability to catalyze a chemical reaction by binding to and stabilizing the transition state intermediate.

Catalytic RNA (ribozyme): A natural or synthetic RNA molecule that cuts an RNA substrate.

Cation: A positively charged ion.

cDNA library: A library composed of complementary copies of cellular mRNAs.

cDNA: DNA synthesized from an RNA template using reverse transcriptase. Complementary DNA to a particular RNA fragment.

Cell pharmacology: Delivery of drugs by medical nanomachines to exact locations in the body.

Cell Repair Machine: A system including nanocomputers and molecular scale sensors and tools, programmed to repair damage to cells and tissues. Molecular and nanoscale machines with sensors, nanocomputers and tools, programmed to detect and repair damage to cells and tissues, which could even report back to and receive instructions from a human doctor if needed.

Cell surgery: In medical nanorobotics, modifying cellular structures using medical nanomachines. Modifying cellular structures using medical nanomachines.

Cell: A membrane-bound unit, typically microns in diameter. All plants and animals are made up of one or more cells (trillions, in the case of human beings). In general, each cell of a multicellular organism contains a nucleus holding all of the genetic information of the organism. (2) A small structural unit, surrounded by a membrane, making up living things.

Cellular oncogene (proto-oncogene): A normal gene that when mutated or improperly expressed contributes to the development of cancer.

Centers of origin: Usually the location in the world where the oldest cultivation of a particular crop has been identified.

Central dogma: Francis Crick's seminal concept that in nature genetic information generally flows from DNA to RNA to protein.

Centrifugation: Separating molecules by size or density using centrifugal forces generated by a spinning rotor.

G forces of several hundred thousand times gravity are generated in ultracentrifugation.

Centromere: The central portion of the chromosome to which the spindle fibers attach during mitotic and meiotic division.

Chemotherapy: A treatment for cancers that involves administering chemicals toxic to malignant cells.

Chloramphenicol: An antibiotic that interferes with protein synthesis.

Chromatid: Each of the two daughter strands of a duplicated chromosome joined at the centromere during mitosis and meiosis.

Chromosome walking: Working from a flanking DNA marker, overlapping clones are successively identified that span a chromosomal region of interest.

Chromosome: A single DNA molecule, a tightly coiled strant of DNA, condensed into a compact structure in vivo by complexing with accessory histones or histone-like proteins. Chromosomes exist in pairs in higher eukaryotes. A linear structure in the nucleus of plants and animals that is visible in light microscopy when stained. The chromosome is a single, long, linear molecule of DNA and associated proteins. Bacteria have a single circular chromosome; other organisms may have many linear chromosomes.

Cistron: A DNA sequence that codes for a specific polypeptide; a gene. See DNA, Gene.

Clone: An exact genetic replica of a specific gene or an entire organism. See Cloning.

Cloning vector: A DNA molecule capable of autonomous replication within the cloning host cell (e.g *E. coli*). The vectors contain restriction enzyme sites for insertion of foreign DNA. Cloning vectors are derived from bacterial plasmids, bacteriophages, or viruses.

Cloning: The mitotic division of a progenitor cell to give rise to a population of identical daughter cells or clones.

Coat protein (capsid): The coating of a protein that enclosed the nucleic acid core of a virus.

Codon: A group of three nucleotides that specifies addition of one of the 20 amino acids during translation of an mRNA into a polypeptide. Strings of codons form genes and strings of genes form chromosomes. Set of three nucleotides that specify a particular amino acid during protein synthesis.

Coenzyme (cofactor): An organic molecule, such as a vitamin, that binds to an enzyme and is required for its catalytic activity.

Colony: A group of identical cells (clones) derived from a single progenitor cell.

Combinatorial materials design: Uses computing power (sometimes together with massive parallel experimentation) to screen many different materials possibilities to optimize properties for specific applications (e.g., catalysts, drugs, optical materials).

Commensalism: The close association of two or more dissimilar organisms where the association is advantageous to one and doesn't affect the other(s). See Parasitism, Symbiosis.

Competency: An ephemeral state, induced by treatment with cold cations, during which bacterial cells are capable of uptaking foreign DNA.

Conjugation: The joining of two bacteria cells when genetic material is transferred from one bacterium to another.

Constitutive promoter: An unregulated promoter that allows for continual transcription of its associated gene.

Contiguous (contig) map: The alignment of sequence data from large, adjacent regions of the genome to produce a continuous nucleotide sequence across a chromosomal region.

Culture: A particular kind of organism growing in a laboratory medium.

Curie temperature (also Curie point) (Tc): The temperature above which a ferromagnetic or ferrimagnetic material becomes paramagnetic. For iron the Curie point is 760oC and for nickel 356oC.

Current [A]: Measure of rate of flow of electric charge: a one-ampere current is a flow of 1 C of charge per second.

Cutoff: Condition in a diode or bipolar junction transistor in which the potential across a p-n junction prevents current flow.

Cyclic AMP (cyclic adenosine monophosphate): A second messenger that regulates many intracellular reactions by transducing signals from extracellular growth factors to cellular metabolic pathways.

Cyclic: A structure is termed cyclic if its covalent bonds form one or more rings.

Cytogenetics: Study that relates the appearance and behavior of chromosomes to genetic phenomenon.

Dalton: A unit of measurement equal to the mass of a hydrogen atom, 1.67 x 10E-24 gram/L (Avogadro's number).

Denature: To induce structural alterations that disrupt the biological activity of a molecule. Often refers to breaking hydrogen bonds between base pairs in double-stranded nucleic acid molecules to produce in single-stranded polynucleotides or altering the secondary and tertiary structure of a protein, destroying its activity.

Density gradient centrifugation: High-speed centrifugation in which molecules "float" at a point where their density equals that in a gradient of cesium chloride or sucrose.

Diabetes: A disease associated with the absence or reduced levels of insulin, a hormone essential for the transport of glucose to cells.

Dideoxynucleotide (didN): A deoxynucleotide that lacks a 3' hydroxyl group, and is thus unable to form a 3'-5' phosphodiester bond necessary for chain elongation. Dideoxynucleotides are used in DNA sequencing and the treatment of viral diseases.

Digest: To cut DNA molecules with one or more restriction endonucleases.

Diploid cell: A cell which contains two copies of each chromosome. See Haploid cell.

Directional cloning: DNA insert and vector molecules are digested with two different restriction enzymes to create noncomplementary sticky ends at either end of each restriction fragment. This allows the insert to be ligated to the vector in a specific orientation and prevents the vector from recircularizing.

DNA: DNA is the recipe for life. DNA is a molecule found in the nucleus of every cell and is made up of 4 nucleotides Adenine (A), Guanine (G), Thymine (T), and Cystosine (C). The order of the nucleotides in the DNA strand holds a code of information for the cell.

DNA (Deoxyribonucleic acid): An organic acid and polymer composed of four nitrogenous bases—adenine, thymine, cytosine, and guanine linked via intervening units of phosphate and the pentose sugar deoxyribose.

DNA is the genetic material of most organisms and usually exists as a double-stranded molecule in which two antiparallel strands are held together by hydrogen bonds between adeninethymine and cytosine-guanine.

DNA (Deoxyriobonucleic Acid): DNA molecules are long chains consisting of four kinds of nucleotides; the order of these nucleotides encodes the information needed to construct protein molecules. These in turn make up much of the molecular machinery of the cell. DNA is the genetic material of cells.

DNA Chip: Also: Gene Chip and DNA Microchip. A purpose built microchip used to identify mutations or alterations in a gene's DNA.

DNA diagnosis: The use of DNA polymorphisms to detect the presence of a disease gene.

DNA diagnostics - miniaturization of: In the areas of sample preparation and assay, it is clear that miniaturization is key. To reduce the size of samples by a factor of 10 or greater, barriers in microfluidics, micromachining, robotics, microchemistry, nucleic acid chemistry, and surface chemistry must be overcome. To implement miniaturized protocols accurately and efficiently, substantial automation of the process will be required. In the development of miniaturized systems, it is essential that the system can be adapted for high levels of parallelization. Miniaturization poses significant technological risks. Currently, there exists no universally accepted precedent for the handling, replication, amplification, or cloning of DNA in nanoliter volumes. Due to the size and charge of the DNA molecule, and the relative instability of many of the enzymes involved in the sample preparation processes, nanoliter and less volumes may pose substantial challenges. In addition, interactions of the biological molecules with the surfaces of the reaction chambers must be minimized. For some methodologies, it is not clear what the optimal sample will be, so substantial improvement in DNA fragmentation technologies or DNA cloning vectors may be required for the ultimate efficient application to diagnostics. Improvements in any of these areas are likely to be of value to other non- DNA based diagnostic applications such as antibody screening protocols and enzyme based diagnostics, because miniaturized robotic or micro- electro mechanical systems developed for DNA could be modified to be used for these purposes.

DNA fingerprint: The unique pattern of DNA fragments identified by Southern hybridization (using a probe that binds to a polymorphic region of DNA) or by polymerase chain reaction (using primers flanking the polymorphic region).

DNA polymorphism: One of two or more alternate forms (alleles) of a chromosomal locus that differ in nucleotide sequence or have variable numbers of repeated nucleotide units.

DNA Probes: A DNA or nucleic acid probe is a short strand of DNA that locates and binds to its complementary sequence in samples containing single strands of DNA or RNA enabling identification of specific sequences. Nucleic acid probe assays exploit the fundamental hybridization reaction that occurs spontaneously between two complementary DNA:DNA or DNA:RNA strands. As in immunoassays, detection of the hybrid requires that the probe be labeled. Various direct and indirect methods have been devised for the detection of the hybrid. Direct labeling involves attaching the label directly to the probe sequence; indirect labeling binds an antibody to the DNA:DNA or DNA:RNA hybrid. As in immunoassays, non-isotopically-labeled probes are preferred over radio-labeled probes primarily because of radiation hazards, disposal problems, and short reagent shelf life. In addition, the factors determining the detection limits of hybridization assays based on labeled probes are similar to those in immunoassays. Therefore, the development of a simple, inexpensive and sensitive direct detection system which eliminates the use of labels is highly desirable.

DNA Sequencing: There are two main classical methods for sequencing DNA: The first method, developed by Allan Maxam and Walter Gilbert, involves chemicals used to cleave the DNA at certain positions, generating a set of fragments that differ by one nucleotide. The second method, developed by Fred Sanger and Alan Coulson, involves enzymatic synthesis of DNA strands that terminate in a modified nucleotide. Analysis of fragments is similar for both methods and involves gel electrophoresis and autoradiography or fluorescence. The enzymatic method has largely replaced the chemical method as the technique of choice, although there are some situations where chemical sequencing can provide data more easily than the enzymatic method.

DNA, recombinant: DNA that has been cut and spliced back together in a new sequence. The DNA may be from one organism or from more than one organism

DNA, repetitive: Fragments of DNA that appear in multiple copies in a single individual.

DNA: A molecule encoding genetic information, found in the cell's nucleus. Deoxyribose Nucleic Acid. DNA is a molecule used in biological systems to store genetic information. It has a long, double stranded structure with the two strands wrapped around each other to form a right handed helix. There are about 10 nucleotide pairs per helical turn. Each strand is composed of a sugar phosphate backbone and attached bases. It is connected to a complementary strand by hydrogen bonding (non- covalent) between paired bases, adenine (A) with thymine (T) and guanine (G) with cytosine (C). DNA is an abbreviation of DeoxyriboNucleic Acid. : It is the genetic 'blueprint' of all life on this planet. : DNA typing is often applied to Human Identification which has also been called DNA Fingerprinting. DNA Fingerprinting has been attributed to Alex Jeffries, who with his colleagues in 1985, discovered a sequence of nucleotides, subsequently termed short tandem repeats (STR) when comparing the DNA sequence of the human á-globin chromosomes isolated from different individuals.

DNA: Deoxyribonucleic acid, the molecule that stores genetic information. Composed of two complementary strands. See base pairs.

Dominant gene: A gene whose phenotype is when it is present in a single copy.

Dominant(-acting) oncogene: A gene that stimulates cell proliferation and contributes to oncogenesis when present in a single copy.

Dominant: An allele is said to be dominant if it expresses its phenotype even in the presence of a recessive allele. See Allele, Phenotype, Recessive.

Dormancy: A period in which a plant does not grow, awaiting necessary environmental conditions such as temperature, moisture, nutrient availability.

Double helix: Describes the coiling of the antiparallel strands of the DNA molecule, resembling a spiral staircase in which the paired bases form the steps and the sugar-phosphate backbones form the rails.

Double-stranded complementary DNA (dscDNA): A duplex DNA molecule copied from a cDNA template.

Downstream: The region extending in a 3' direction from a gene.

E: coli (Escherichia coli): A common intestinal bacterium that is widely used in genetic engineering as a host for a cloning vector. Some strains of E. coli are important foodborne pathogens. Lab strains are of the harmless variety.

Ecology: The study of the interactions of organisms with their environment and with each other.

Ecosystem: The organisms in a plant population and the biotic and abiotic factors which impact on them. See abiotic factors; Biotic factors.

Electrophoresis: The technique of separating charged molecules in a matrix to which is applied an electrical field.

Electroporation: A method for transforrning DNA, especially useful for plant cells, in which high voltage pulses of electricity are used to open pores in cell membranes, through which foreign DNA can pass.

Electroporation: Transformation technique that uses electric fields to temporarily increase permeability of cells to foreign DNA.

Emergence: A complex whole created by simple parts, as in the brain where billions of neurons work individually, but collectively make up our consciousness and give us the ability to think, rationalize, and create.

Emulation: An absolutely precise simulation of something, so exact that it is equivalent to the original (for example, many computers emulate obsolete computers to run their programs). [AS] The Star Trek replicator is an example.

Enabling science and technologies: Areas of research

relevant to a particular goal, such as nanotechnology. Also, technology that "enables" other technology to advance, such as the transistor enabled the computer chip revolution, as did photolithography.

Encapsidation: Process by which a virus' nucleic acid is enclosed in a capsid. See Coat protein.

Endocrine cell: Specialized animal cell that secretes a hormone into the blood; usually part of a gland, such as the thyroid or pituitary gland.

Endocytosis: Uptake of material into a cell by an invagination of the plasma membrane and its internalization in a membrane-bounded vesicle.

Endoplasmic reticulum (ER): Labyrinthine, membrane-bounded compartment in the cytoplasm of eucaryotic cells, where lipids are synthesized and membrane-bound proteins are made.

Endosome: Membrane-bounded organelle in animal cells that carries materials newly ingested by endocytosis and passes many of them on to lysosomes for degradation.

Endothelium: Single sheet of highly flattened cells (endothelial cells) that forms the lining of all blood vessels. Regulates exchanges between the bloodstream and surrounding tissues and is usually surrounded by a basal lamina.

Energy [J]: Capacity for performing work or to cause heat flow. Like work itself, it is measured in Joules.

Environmental Protection Agency (EPA): The U.S. regulatory agency for biotechnology of microbes. The major laws under which the agency has regulatory powers are the Federal Insecticide, Fungicide, and Rodenticide Act (FIFRA); and the Toxic Substances Control Act (TSCA).

Enzymes: Proteins that control the various steps in all chemical reactions.

EPROM: Electrically Programmable Read-Only Memory — nonvolatile memory device.

Escherichia coli: A commensal bacterium inhabiting the human colon that is widely used in biology, both as a simple model of cell biochemical function and as a host for molecular cloning experiments.

Ethidium bromide: A fluorescent dye used to stain DNA and RNA. The dye fluoresces when exposed to UV light.

Eukaryote: An organism whose cells possess a nucleus and other membrane-bound vesicles, including all members of the protist, fungi, plant and animal kingdoms; and excluding viruses, bacteria, and blue-green algae. See Prokaryote.

Eurisko: A computer program developed by Professor Douglas Lenat which is able to apply heuristic rules for performing various tasks, including the invention of new heuristic rules.

Event: An "event" in genetic engineering is the insertion of a particular piece of foreign DNA into the chromosome of the recipient. Insertion occurs in random locations, so each event is unique. The event can affect how a gene is expressed in the organism. Once an event occurs, the transgene can be passed to the next generation as a normally inherited gene.

Evolution: The long-term process through which a population of organisms accumulats genetic changes that enable its members to successfully adapt to environmental conditions and to better exploit food resources.

Exon: A DNA sequence that is ultimately translated into protein.

Expectation value: The number of different alignents with scores equivalent to or better than S that are expected to occur in a database search by chance. The lower the E value, the more significant the score.

Express: To translate a gene's message into a molecular product.

Face-centered cubic: A crystal structure found in common elemental metals also FCC. Within the cubic unit cell, atoms are located at all corner and face-centered positions.

Fail-stop: Describes a component or subsystem that, in the event of a failure, produces no output (e.g., of material or data) rather than producing a damaged or incorrect output.

Fermi energy: The energy level in a solid at which the probability of finding an electron is 1/2. For a metal, the energy corresponding to the highest filled electron state in the valence band at 0 K.

Field-Effect Transistor (FET): Field-Effect Transistor — semiconductor device whose insulated gate electrode controls current flow.

FIFRA: The Federal Insecticide, Fungicide, and Rodenticide Act. See Environmental Protection Agency.

Fingerprint: A set of molecular markers sufficiently diverse to identify particular individuals with reasonable certainty.

Flanking region: The DNA sequences extending on either side of a specific locus or gene.

Food and Drug Administration (FDA): The U.S. agency responsible for regulation of biotechnology food products.

The major laws under which the agency has regulatory powers include the Food, Drug, and Cosmetic Act; and the Public Health Service Act.

Fusion gene: A hybrid gene created by joining portions of two different genes (to produce a new protein) or by joining a gene to a different promoter (to alter or regulate gene transcription).

Gamete: A haploid sex cell, egg or sperm, that contains a single copy of each chromosome.

Ganglioside: Any glycolipid having one or more sialic acid residues in its structure. Found in the plasma membrane of eucaryotic cells and especially abundant in nerve cells.

Gel, agarose: A substance used to separate DNA or RNA fragments by size. Used for southern and northern blots.

Gel, polyacrylamide: A substance used to separate proteins by size. Used for western blots.

GEM: A genetically engineered microorganism.

Gene amplification: The presence of multiple genes. Amplification is one mechanism through which proto-oncogenes are activated in malignant cells.

Gene cloning: The process of synthesizing multiple copies of a particular DNA sequence using a bacteria cell or another organism as a host. See DNA, Host.

Gene expression: The process of producing a protein from its DNA- and mRNA-coding sequences.

Gene flow: The exchange of genes between different but (usually) related populations.

Gene frequency: The percentage of a given allele in a population of organisms. See Allele.

Gene gun: Transformation technique that uses accelerated particles coated with DNA to introduce foreign DNA into recipient plant.

Gene insertion: The addition of one or more copies of a normal gene into a defective chromosome.

Gene linkage: The hereditary association of genes located on the same chromosome.

Gene modification: The chemical repair of a gene's defective DNA sequence. See DNA.

Gene pool: The totality of all alleles of all genes of all individuals in a particular population.

Gene therapy: The introduction of new genes into individuals to cure diseases or genetic abnormalities.

Gene translocation: The movement of a gene fragment from one chromosomal location to another, which often alters or abolishes expression.

Gene: The basic unit of inheritance. A segment of DNA that codes for a particular protein. A locus on a chromosome that encodes a specific protein or several related proteins. It is considered the functional unit of heredity.

Genetic assimilation: Eventual extinction of a natural species as massive pollen flow occurs from another related species and the older crop becomes more like the new crop. See Gene flow.

Genetic code: The three-letter code that translates nucleic acid sequence into protein sequence. The relationships between the nucleotide base-pair triplets of a messenger RNA molecule and the 20 amino acids that are the building blocks of proteins. See Base pair, Nucleic acid, Nucleotide. The genetic information in DNA is encoded with four different nucleotide bases: A, C, G, and T. A set of three consecutive nucleotide bases constitutes a codon. A codon specifies a particular amino acid that is added during synthesis of a protein.

Genetic disease: A disease that has its origin in changes to the genetic material, DNA. Usually refers to diseases that are inherited in a Mendelian fashion, although noninherited forms of cancer also result from DNA mutation.

Genetic drift: Random variation in gene frequency from one generation to another.

Genetic engineering: Genetic engineering is the process of manually adding new DNA on a molecular level with the goal of adding one or more new traits that are not already found in that organism. For the purposes of this web site, genetic engineering includes recombinant DNA and gene splicing technologies.

Genetic linkage map: A linear map of the relative positions of genes along a chromosome. Distances are established by linkage analysis, which determines the frequency at which two gene loci become separated during chromosomal recombination. (

Genetic marker: A gene or group of genes used to "mark" or track the action of microbes.

Genetic modification: The selective and purposeful alteration of genes by humans. Genetic modification includes methods of incorporating new genes or traits into an organism through genetic engineering, mutagenesis and traditional breeding methods.

Genome: The genetic complement contained in the chromosomes of a given organism, usually the haploid chromosome state. The full chromosome set containing all the genes of a particular individual.

Genomic library: A library composed of fragments of genomic DNA.

Genomics: The study of the structure and function of genomes. Genomics usually involves high speed sequencing of the DNA and computer searches for sequences that code for genes.

Genotype: The sum total of the genetic information of an organism including the linkage relationships between genes. The genotype, modified by environment, determines the phenotype.

Genus: A category including closely related species. Interbreeding between organisms within the same category can occur.

Germ cell (germ line) gene therapy: The repair or replacement of a defective gene within the gamete-forming tissues, which produces a heritable change in an organism's genetic constitution.

GMO: Genetically modified organism. An organism that has incorporated a functional foreign gene through recombinant DNA technology. The novel gene exists in all of its cells and is passed through to progeny. Same as transgenic.

Green revolution: Advances in genetics, petrochemicals, and machinery that culminated in a dramatic increase in crop productivity during the third quarter of the 20th century.

Growth curve: See Growth phase.

Growth factor: A serum protein that stimulates cell division when it binds to its cell-surface receptor.

Growth phase (curve): The characteristic periods in the growth of a bacterial culture, as indicated by the shape of a graph of viable cell number versus time.

GUI: Graphical User Interface — hardware, software, and firmware that produces the display on modern personal computers.

Guy Fawkes Scenario: If nanotechnology becomes widely available, it might become trivial for anyone to committ acts of terrorism (such as making nanomachines build a large amount of explosives under government buildings a la Guy Fawkes). This would either force strict control over nanotechnology (hard) or a decentralized mode of organization.

Hair cell: Specialized sensory epithelial cell in the ear with bundles of giant microvilli (stereocilia) protruding from its apical surface. Sound vibrations tilt the stereocilia, evoking an electrical change in the hair cell, which thus acts as a sound detector.

Hall effect: The phenomenon whereby a force is applied to a moving electron or hole by a magnetic field that is applied perpendicular to the direction of motion. The force direction is perpendicular to both the magnetic field and the particle motion directions.

Haploid cell: A cell containing only one set, or half the usual (diploid) number, of chromosomes.

Heme: An iron complex. Cyclic organic molecule containing an iron atom that carries oxygen in hemoglobin and carries an electron in cytochromes.

Hemophilia: An X-linked recessive genetic disease, caused by a mutation in the gene for clotting factor VIII (hemophilia A) or clotting factor IX (hemophilia B), which leads to abnormal blood clotting.

Herbicide resistance: Herbicide resistance is when plants become resistant to the negative effects of a particular herbicide formulation or are genetically engineered in order to have resistance to a particular herbicide.

Herbicide-resistant crops can be sprayed with a particular herbicide, killing weeds growing in and around the crop plants, without being damaged.

Herbicide: Any substance that is toxic to plants; usually used to kill specific unwanted plants.

Heterochromatin: Dark-stained regions of chromosomes thought to be for the most part genetically inactive.

Heteroduplex: A double-stranded DNA molecule or DNA-RNA hybrid, where each strand is of a different origin.

Heterogeneous nuclear RNA (hnRNA): The name originally given to large RNA molecules found in the nucleus, which are now known to be unedited mRNA transcripts, or pre-mRNAs.

Homologous chromosomes: Chromosomes that have the same linear arrangement of genes—a pair of matching chromosomes in a diploid organism. See Chromosomes.

Homologous recombination: The exchange of DNA fragments between two DNA molecules or chromatids of paired chromosomes (during crossing over) at the site of identical nucleotide sequences.

Homozygote: An organism whose genotype is characterized by two identical alleles of a gene. See Allele, Genotype.

Host: An organism that contains another organism.

Human Genome Project: A project coordinated by the National Institutes of Health (NIH) and the Department of Energy (DOE) to determine the entire nucleotide sequence of the human chromosomes.

Human growth hormone (HGH, somatotrophin): A protein produced in the pituitary gland that stimulates the liver to produce somatomedins, which stimulate growth of bone and muscle.

Hybrid: The offspring of two parents differing in at least one genetic characteristic (trait). Also, a heteroduplex DNA or DNA-RNA molecule.

Hybridization, nucleic acid: Complementary strands of DNA or RNA will spontaneously match up with each other and bond together under the right conditions. This is called nucleic acid hybridization and is used when probing southern and northern blots.

Hybridization: The hydrogen bonding of complementary DNA and/or RNA sequences to form a duplex molecule.

Hybridoma: A hybrid cell, composed of a B lymphocyte fused to a tumor cell, which grows indefinitely in tissue culture and is selected for the secretion of a specific antibody of interest.

Hydrogen bond: A relatively weak bond formed between a hydrogen atom (which is covalently bound to a nitrogen or oxygen atom) and a nitrogen or oxygen with an unshared electron pair.

Hydrolysis: A reaction in which a molecule of water is added at the site of cleavage of a molecule to two products.

Imaging Contrast Agent: A molecule or molecular complex that increases the intensity of the signal detected by an imaging technique, including MRI and ultrasound. An MRI contrast agent, for example, might contain gadolinium attached to a targeting antibody. The antibody would bind to a specific target—a metastatic melanoma cell, for example – while the gadolinium would increase the magnetic signal detected by the MRI scanner.

Immortalization: Production of a cell line capable of an unlimited number of cell divisions. Can be the result of a chemical or viral transformation or of fusion with cells of a tumor line.

Immortalizing oncogene: A gene that upon transfection enables a primary cell to grow indefinitely in culture.

In situ: Refers to performing assays or manipulations with intact tissues.

In vivo: Refers to biological processes that take place within a living organism or cell.

Incomplete dominance: A condition where a heterozygous off- spring has a phenotype that is distinctly different from, and intermediate to, the parental phenotypes. See Heterozygote, Phenotype.

Initiation codon: The mRNA sequence AUG, coding for methionine, which initiates translation of mRNA.

Inositol lipid: A membrane-anchored phospholipid that transduces hormonal signals by stimulating the release of any of several chemical messengers.

insect resistance management: A set of strategies to reduce the frequency of resistance in crop pests. These strategies are widely referred to as Insect Resistance Management.

Insertion mutations: Changes in the base sequence of a DNA molecule resulting from the random integration of DNA from another source. See DNA, Mutation.

Insulator: A substance, object or material that does not conduct electricity. Material that conducts electricity very poorly.

Insulin: A peptide hormone secreted by the islets of Langerhans of the pancreas that regulates the level of sugar in the blood.

Integrated Circuit (IC): An electronic circuit consisting of many interconnected devices on one piece of semiconductor, typically into 10 millimeters on a side. ICs are the major building blocks of today's computers.

Semiconductor circuit, typically on a very small silicon chip, containing microfabricated transistors, diodes, resistors, capacitors, etc.

Interferon: A family of small proteins that stimulate viral resistance in cells.

Intergenic regions: DNA sequences located between genes that comprise a large percentage of the human genome with no known function.

Introgression: Backcrossing of hybrids of two plant populations to introduce new genes into a wild population.

Intron: A noncoding DNA sequence within a gene that is initially transcribed into messenger RNA but is later snipped out. See Coding, DNA, Messenger RNA, Transcription.

Invasiveness: Ability of a plant to spread beyond its introduction site and become established in new locations where it may provide a deliterious effect on organisms already existing there.

Ion: A charged particle.

Isotope: One of two or more forms of an element that have the same number of protons (atomic number) but differing numbers of neutrons (mass numbers). Radioactive isotopes are commonly used to make DNA probes and metabolic tracers.

Joining (J) segment: A small DNA segment that links genes to yield a functional gene encoding an immunogobulin.

Jupiter-Brain: A posthuman being of extremely high computational power and size. This is the archetypal concentrated intelligence. The term originated due to an idea by Keith Henson that nanomachines could be used to turn the mass of Jupiter into computers running an upgraded version of himself.

Kanamycin: An antibiotic of the aminoglycoside family that poisons translation by binding to the ribosomes.

Karyotype: All of the chromosomes in a cell or an individual organism, visible through a microsope during cell division.

Kb (kilobase): The number of base pairs (in thousands) that denote the size of a nucleic acid fragment.

Kilohertz (kHz): One thousand cycles per second.

Kilojoule: Standard unit of energy equal to 1000 joules, or 0.24 kilocalories.

Lag phase: The initial growth phase, during which cell number remains relatively constant prior to rapid growth. See growth phase.

Lane: On a gel or a blot, a lane belongs to one sample. It contains fragments of different sizes that were separated by gel electrophoresis. See band, blot.

Lawn: A uniform and uninterrupted laver of bacterial growth, in which individual colonies cannot be observed.

Lectin: Protein that binds tightly to a specific sugar. Abundant lectins derived from plant seeds are often used as affinity reagents to purify glycoproteins or to detect them on the surface of cells.

Legume: A member of the pea family that possesses root nodules containing nitrogen-fixing bacteria.

Library, DNA: A large set of clones of DNA fragments from a particular organism. DNA libraries are usually maintained in *E. coli* or in bacteriophages.

Life (lifetime): The length of time the sensor can be used before its performance changes.

Life Science Applications of Deep Red T2-MP EviTag Quantum Dots: The new T2-MP EviTags are available in the deep red wavelengths used in life science research – Macoun Red (660nm) and Jonamac Red (680nm) –

with carboxyl, amine, biotin and non-functionalized surfaces. Having a biologically compatible quantum dot label like the T2-MP EviTags allows further exploration and development of in vivo assays, following cells inside living organism to get a new understanding of life. In particular, it offers the possibility to create new classes of assays for use where current testing materials could produce cytotoxicity or cell death. Potential applications include live cell and even whole animal imaging, blood cancer assays, and numerous other applications.

Ligand: In protein chemistry, a small molecule that is (or can be) bound by a larger molecule is termed a ligand. In organometallic chemistry, a moiety bonded to a central metal atom is also termed a ligand; the latter definition is more common in general chemistry.

Ligase (DNA ligase): An enzyme that catalyzes a condensation reaction that links two DNA molecules via the formation of a phosphodiester bond between the 3' hydroxyl and 5' phosphate of adjacent nucleotides.

Ligate: The process of joining two or more DNA fragments.

Ligation: Joining two fragments of DNA end to end. See sticky ends.

Light Emitting Diodes (LEDs): Light Emitting Diodes work on a completely different concept. Traditionally LEDs are created from two semiconductors. By running current in one direction across the semiconductor the LED emits light of a particular frequency (hence a particular color) depending on the physical characteristics of the semiconductor used. The semiconductor is covered with a piece of plastic that focuses the light and increases the brightness. These semiconductors are very durable, there is no filament, they donít require much power, theyíre brighter and they last a long time. By densely packing red, blue and green LEDs next to each other on a substrate one can create a display.

Lineage: A chart that traces the flow of genetic information from generation to generation.

Linkage: The frequency of coinheritance of a pair of genes and/or genetic markers, which provides a measure of their physical proximity to one another on a chromosome. Measures the physical distance between two genes. Genes that are close together are unlikely to segregate in a sexual cross. Distant genes segregate independently are are then said to be unlinked.

Linked genes/markers: Genes and/or markers that are so closely associated on the chromosome that they are coinherited in 80% or more of cases.

Linker: A short, double-stranded oligonucleotide containing a restriction endonuclease recognition site, which is ligated to the ends of a DNA fragment.

Liposome: A type of nanoparticle made of lipids, or fat molecules, surrounding a water core. Liposomes, several of which are widely used to treat infectious diseases and cancer, were the first type of nanoparticle to be used to create therapeutic agents with novel characteristics. Membrane-bound vesicles constructed in the laboratory to transport biological molecules.

Liquid Crystal Display: LCD is the predominant technology used in flat panel displays. The principle that makes the display work is this: A crystalís alignment can be altered with an electric current. If the crystal is lined up one way ñ it will allow the light waves to pass through a polarized filter, but if the electric current alters the crystalís alignment, it will guide light so that the polarized filter blocks the light. By densely packing red, blue and green light emitting crystals next to each other on a sheet (ìcalled a substrateî), one can create a full color display. The great thing about LCD is that the crystals can be packed together closely, allowing for a higher-resolution, finer-detail display. The con is that LCDs are somewhat fragile, require a lot of power and are relatively less bright. Liquid Crystal Display — display device employing light source and electrically alterable optically active thin film.

Locus (plural = loci): A specific location or site on a chromosome.

Logarithmic phase (log or exponential growth phase): The steepest slope of the growth curve—the phase of vigorous growth during which cell number doubles every 20-30 minutes.

Lysogen: A bacterial cell whose chromosome contains integrated viral DNA.

Lysogenic: A type or phase of the virus life cycle during which the virus integrates into the host chromosome of the infected cell, often remaining essentially dormant for some period of time. See Lysogen.

Lytic: A phase of the virus life cycle during which the virus replicates within the host cell, releasing a new generation of viruses when the infected cell lyses.

Machine-phase chemistry: The chemistry of systems in which all potentially reactive moieties follow controlled trajectories (e.g., guided by molecular machines working in vacuum).

Macroscale: Larger than nanoscale; often implies a design that humans can directly interact with; too large to be built by a single assembler (one cubic micron of diamond contains 176 billion atoms).

Magnetization (M): The total magnetic moment per unit volume of material. Also, a measure of the contribution to the magnetic flux by some material within an H field. The magnitude of M is proportional to the applied field as: M = ÷m × H, with ÷m the magnetic susceptibility.

Magnetostrictive material: A material that changes dimension in the presence of a magnetic field or generates a magnetic field when mechanically deformed.

Malignant: Having the properties of cancerous growth.

Manufacturing: The ability to make products, in this case ranging from clothing, to electronics, to medical devices, to books, to building materials, and much more.

Meat Machine: AKA Cabinet Beast. A box containing assemblers and raw material, within which is formed meat [or whatever else it was programmed to make].

Mechanical: Pertaining to the positions and motions of atoms, as defined by the positions of their nuclei; see electronic. A purely mechanical device can be described in terms of atomic positions and motions without reference to electronic properties, save through their effect on the potential energy function.

Mechanochemistry: Chemistry accomplished by mechanical systems directly controlling the reactant molecules; the formation or breaking of chemical bonds under direct mechanical control. In this volume, the chemistry of processes in which mechanical systems operating with atomic-scale precision either guide, drive, or are driven by chemical transformations. In general usage, the chemistry of processes in which energy is converted from mechanical to chemical form, or vice versa.

Mechanosynthesis: Chemical synthesis controlled by mechanical systems operating with atomic-scale precision, enabling direct positional selection of reaction sites; synthetic applications of mechanochemistry. Suitable mechanical systems include AFM mechanisms, molecular manipulators, and molecular mill systems. Processes that fall outside the intended scope of this definition include reactions guided by the incorporation of reactive moieties into a shared covalent framework (i.e., conventional intramolecular reactions), or by the binding of reagents to enzymes or enzymelike catalysts. Molecular tools with chemically specific tip structures can be used, sequentially, to modify a work piece and build a wide range of molecular structures.

Mechatronics: The study of the melding of AI and electromechanical machines to make machines that are greater than the sum of their parts. The synergistic combination of precision mechanical engineering with electronic control.

Megabase cloning: The cloning of very large DNA fragments.

Megabyte (MB): 220 (= 1,048,576, or about one million) bytes of information.

Megahertz (Mhz): One million cycles per second

Meiosis: The reduction division process by which haploid gametes and spores are formed, consisting of a single duplication of the genetic material followed by two mitotic divisions.

Meme: An idea that replicates through a society as it is propagated through person-to-person interaction, both direct and indirect. Memetics is a field of study that focuses on memes' role in the evolution of a culture. Self-reproducing idea or other information pattern which is propagated in ways similar to that of a gene. [Richard Dawkins, 1976]

Messenger RNA (mRNA): The class of RNA molecules that copies the genetic information from DNA, in the nucleus, and carries it to ribosomes, in the cytoplasm, where it is translated into protein.

Metabolism: The biochemical processes that sustain a living cell or organism.

Metallothionein: A protective protein that binds heavy metals, such as cadmium and lead.

Micellar nanocontainers: Block copolymer micelles are water- soluble biocompatible nanocontainers with great potential for delivering hydrophobic drugs. An understanding of their cellular distribution is essential to achieving selective delivery of drugs at the subcellular level. R. Savic, L. Luo, A. Eisenberg, D. Maysinger, Micellar nanocontainers distribute to defined cytoplasmic organelles Science 300 (5619): 615- 618, Apr. 25, 2003

Micelles: Micelles are small, spherical structures composed of molecules that attract one another to reduce surface tension. The head of the molecule is hydrophilic, meaning it likes water, while the interior portion is hydrophobic, meaning it avoids water.

Microbial mats (biofilms): Layered groups or communities of microbial populations.

Microelectronics: Narrower terms: MEMS, nanoelectronics, optoelectronics, SED Single Electron Devices. Related terms: molecular electronics, semiconductors

Microencapsulation: Individually encapsulated small particles.

Microinjection: A means to introduce a solution of DNA, protein, or other soluble material into a cell using a fine microcapillary pipet.

Microspectrophotometry: Analytical technique for

studying substances present at enzyme concentrations in single cells, in situ, by measuring light absorption. Light from a tungsten strip lamp or xenon arc dispersed by a grating monochromator illuminates the optical system of a microscope. The absorbance of light is measured (in nanometers) by comparing the difference between the image of the sample and a reference image. MeSH 1990

Microspheres: Microspheres are generally defined as small spheres made of any material and sized from about 0.5 μm to 100 μm. Similar, but smaller spheres sized 10 to 500 nm are called *nanospheres.* Ideally, microspheres are completely spherical and homogeneous in size, although less perfect particles ·are often termed microspheres as well. Depending on the preparation method and material used, microspheres show a typical size distribution which often deviates from the mono-sized ideal.

Microstructure: In material engineering the structural features of a material such as grain boundaries, grain size and structure, subject to observation under a microscope, selective etching etc. In MEMS microstructure unfortunately is also used to designate a micromachined feature. The last decade has seen rapid developments in the fabrication, characterization and conceptual understanding of synthetic microstructures in many different material systems including silicon, III-V and II-VI semiconductors, metals, ceramics and organics. The objective of this journal [Superlattices and Microstructures] is to provide a common interdisciplinary platform for the publication of the latest research results on all such "nanostructures" with dimensions in the range of 1-100 nm; the unifying theme here being the dimensions of these artificial structures rather than the material system in which they are fabricated.

Microsystem: A microscale machine that can sense information from the environment and act accordingly. Outside the U.S., it can also refer to microelectromechanical systems (MEMS). [smalltimes glossary, 2002]

MicroTas, micro Total Analysis Systems, uTAS: Although initial research dates back to the early 1970's, the field of micro- TAS formally started in 1990, when Manz et al described the possibility of creating microsystems that would take care of many or all the traditional analytical steps involved in a biochemical analysis (sample introduction, handling, extraction, purification, concentration, filtration, analysis, detection) .... Micro- TAS offer many advantages over traditional analysis systems. Low power consumption and small reaction volumes, faster analysis, ultrasensitive detection, and minimal human intervention are key parameters in the development of micro- TAS. Most biochemical reactions take place in liquid environments. Hence, the development of MicroTAS is intrinsically linked to the design of liquid handling micro- devices. [Biomedical Applications Group (GAB) Centro Nacional de Microelectronica (CNM- IMB) Bellaterra, Spain, 2000]

Microtubule: A hollow cylindrical tube used for the transport of materials inside cells.

Miller indices: A set of 3 integers (4 for hexagonal) that designate crystallographic planes, as determined from reciprocals of fractional axial intercepts.

Millimeter: One thousandth of a meter, or about 1/26 of an inch.

Mitosis: The replication of a cell to form two daughter cells with identical sets of chromosomes.

Molecular - Nanotechnology: The technology of precisely-constructed molecular-scale machines; from nanometer: a billionth of a meter. For a good introduction, see What is Nanotechnology? or Ralph C. Merkle's nanotechnology page.

Molecular assembler: A general-purpose device for molecular manufacturing, able to guide chemical reactions by positioning individual molecules to atomic accuracy (e.g. mechanosynthesis) and to construct a wide range of useful and stable molecular structures according to precise specifications. Also known as an assembler, a molecular assembler is a molecular machine that can build a molecular structure from its component building blocks. [ZY]

Molecular Beam Epitaxy: [MBE] Process used to make compound (multi-layer) semiconductors. Consists of depositing alternating layers of materials, layer by layer, one type after another (such as the semiconductors gallium arsenide and aluminum gallium arsenide).

Molecular biology: The study of the biochemical and molecular interactions within living cells.

Molecular cloning: The biological amplification of a specific DNA sequence through mitotic division of a host cell into which it has been transformed or transfected.

Molecular genetics: The study of the flow and regulation of genetic information between DNA, RNA, and protein molecules.

Molecular markers: Genetic traits which are detectable on gels or blots and can be used to construct genetic maps. Molecular markers usually have no known function. RFLPs are molecular markers.

Molecular mechanics models: Many of the properties of molecular systems are determined by the molecular potential energy function. Molecular mechanics models approximate this function as a sum of 2-atom, 3-atom, and 4-atom terms, each determined by the geometries and bonds of the component atoms. The 2-atom and 3-atom terms describing bonded interactions roughly correspond to linear springs.

Molecular medicine: A variety of pharmaceutical techniques and gene therapies that address specific molecular diseases or molecular defects in biological systems. A variety of pharmaceutical techniques and therapies in use today. Studying molecules as they relate to health and disease, and manipulating those molecules to improve the diagnosis, prevention, and treatment of disease.

Molecular mill: A mechanochemical processing system characterized by limited motions and repetitive operations without programmable flexibility.

Molecular motors: Protein based machines that are involved in or cause movement such as the rotary devices (flagellar motor and the F1 ATPase) or the devices whose movement is directed along cytoskeletal filaments (myosin, kinesin and dynein motor families). [MeSH, 1999]

Molecular Nanoscience: An emerging interdisciplinary field that combines the study of molecular/ biomolecular systems with the science and technology of nanoscale structures and systems. The potential applications for this research are very broad and include such possibilities as 1) the use of biomolecules and cellular systems to self- assemble nanoelectronic circuitry and other nanoscale structures and 2) the use of lamellar host frameworks containing nano pores that can be tailored to include guest molecules for separation of chemicals for pharmaceutical and other applications. Molecular

Nanoscience Alliance for Interdisciplinary Studies and Activities, Univ. of Minnesota, US,

Molecular Wire: A molecular wire—the simplest electronic component—is a quasi-one-dimensional molecule that can transport charge carriers (electrons or holes) between its ends. [Michael D Ward]

Molecularly imprinted polymers MIPs: A new class of materials that have artificially created receptor structures. Since their discovery in 1972, MIPs have attracted considerable interest from scientists and 3engineers involved with the development of chromatographic absorbents, membranes, sensors and enzyme and receptor mimics.

Monoclonal antibodies: Immunoglobulin molecules of single- epitope specificity that are secreted by a clone of B cells.

Monoculture: The agricultural practice of cultivating crops consisting of genetically similar organisms.

Monogenic: Controlled by or associated with a single gene.

Multi-locus probe: A probe that hybridizes to a number of different sites in the genome of an organism.

Mutagen: Any agent or process that can cause mutations. See Mutation.

Mutation: An alteration in DNA structure or sequence of a gene. A spontaneous or induced genetic change in the DNA of an organism. Most mutations are undesirable, but a few are useful such as dwarfing genes in cereal crops that allow the plants to stand better.

Mycorrhizae: Fungi that form symbiotic relationships with roots of more developed plants.

Nanobiology: Many fundamental biological functions are carried out by molecular machineries that have the sizes of 1-100 nm. You find many examples in molecular biology and cell biology: single enzymes, transcription complex, ribosome, transport complex, nuclear pore, and so on. To understand the functions of these machineries, one has to describe their movements, changes in their shapes, and their localization. This means the mechanistic study is equivalent to dynamic morphology at this level of size, making a new field that merges mechanistic biology and morphology. The emergence of nanobiology depended on the invention of nano- technology: scanning probe microscopy, modern optical techniques, and micro- manipulating techniques. This concept of nanobiology was first proposed in a group study named "Biological Nano- Mechanisms", which was supported by Japanese Agency of Science and Technology (1992-1997). [National Institute of Genetics, Japan]

Nanobiotechnology: An emerging area of scientific and technological opportunity. Nanobiotechnology applies the tools and processes of nano/ microfabrication to build devices for studying biosystems. Researchers also learn from biology how to create better micro- nanoscale devices. The Nanobiotechnology Center (NBTC), a National Science Foundation, Science and Technology Center is characterized by its highly interdisciplinary nature and features a close collaboration between life scientists, physical scientists, and engineers. [NBTC, NanoBiotechnology Center, Cornell Univ. US, 2002] Use of nanotechnology[ies] in the life sciences. These may include, but are not limited to, therapeutics, medical devices/implants, biosensors, and tools for the development of drugs. Applying the tools and processes of MNT to build devices for studying biosystems, in order to learn from biology how to create better nanoscale devices. Should hasten the creation of useful micro devices that mimic living biological systems.

Nanocentrifuge: In medical nanorobotics, a proposed nanodevice that can spin materials at very high speed, imparting rotational accelerations of up to one trillion gravities (g's), thus permitting rapid sortation.

Nanoscience: Nanoscience is the science of nanoscale phenomena. This is a very complicated area, since Quantum Mechanics is often needed to describe the physical properties of atoms and molecules properly. What makes it even more complicated is that nanostructured materials can still be quite a challenge to describe Quantum Mechanically, since the equations are feindishly difficult—this is where supercomputers come in to make the job approachable. The scientific discipline seeking to increase our knowledge and understanding of nanoscale phenomena, i.e. science on the scale of 0.1 nm to 100 nm.

Nanosensor: A chemical or physical sensor constructed using nanoscale components, usually microscopic or submicroscopic in size.

Nanosources: sources that emit light from nanometre-scale volumes.

Nanotech: Slang for nanotechnology.

National Institutions of Health (NIH): A nonregulatory agency which has oversight of research activities that the agency funds.

National Science Foundation (NSF): A nonregulatory agency which has oversight of biotechnology research activities that the agency funds.

Natural selection: The differential survival and reproduction of organisms with genetic characteristics that enable them to better utilize environmental resources.

Neophile: One who welcomes the future and who enjoys change and evolution.

Nick translation: A procedure for making a DNA probe in which a DNA fragment is treated with DNase to produce single-stranded nicks, followed by incorporation of radioactive nucleotides from the nicked sites by DNA polymerase I.

Nicked circle (relaxed circle): During extraction of plasmid DNA from the bacterial cell, one strand of the DNA becomes nicked. This relaxes the torsional strain needed to maintain supercoiling, producing the familiar form of plasmid.

Nitrocellulose: A membrane used to immobilize DNA, RNA, or protein, which can then be probed with a labeled sequence or antibody.

Nitrogen fixation: The conversion of atmospheric nitrogen to biologically usable nitrates.

Nitrogenous bases: The purines (adenine and guanine) and pyrimidines (thymine, cytosine, and uracil) that comprise DNA and RNA molecules.

Nodule: The enlargement or swelling on roots of nitrogen-fixing plants. The nodules contain symbiotic nitrogen-fixing bacteria. See Nitrogen fixation.

Nontarget organism: An organism which is affected by an interaction for which it was not the intended recipient.

Northern hybridization: (Northern blotting). A procedure in which RNA fragments are transferred from an agarose gel to a nitrocellulose filter, where the RNA is then hybridized to a radioactive probe.

Novel foods: Novel foods are products that have never been used as a food; foods which result from a process that has not previously been used for food; or, foods that have been modified by genetic manipulation. This last category of foods have been described as genetically modified foods (often referred to as GM foods, genetically engineered foods or biotechnology-derived foods). *Definition from Health Canada Web site*

Nuclease: Any enzyme that cuts nucleic acids. See restriction enzyme. A class of enzymes that degrades DNA and/or RNA molecules by cleaving the phosphodiester bonds that link adjacent nucleotides. In deoxyribonuclease (DNase), the substrate is DNA. In endonuclease, it cleaves at internal sites in the substrate molecule. Exonuclease progressively cleaves from the end of the substrate molecule. In ribonuclease (RNase), the substrate is RNA. In the S1 nuclease, the substrate is single-stranded DNA or RNA.

Nucleic acids: The two nucleic acids, deoxyribonucleic acid (DNA) and ribonucleic acid (RNA), are made up of long chains of molecules called nucleotides. See DNA, RNA, Nucleotides.

Nuclein: The term used by Friedrich Miescher to describe the nuclear material he discovered in 1869, which today is known as DNA.

Nucleoside analog: A synthetic molecule that resembles a naturally occuring nucleoside, but that lacks a bond site needed to link it to an adjacent nucleotide.

Nucleotide: A building block of DNA and RNA, consisting of a nitrogenous base, a five-carbon sugar, and a phosphate group. Together, the nucleotides form codons, which when strung together form genes, which in turn link to form chromosomes.

Nucleus: The membrane-bound region of a eukaryotic cell that contains the chromosomes. A cellular organelle in plants and animals that contains the chromosomes which in turn are composed of DNA plus protein.

Occupational Safety and Health Administration (OSHA): One of the U.S. agencies responsible for regulation of biotechnology. The major law under which the agency has regulatory powers is the Occupational Safety and Health Act.

Oligonucleotide: A DNA polymer composed of only a few nucleotides.

Oncogene: A gene that contributes to cancer formation when mutated or inappropriately expressed.

Oncogenesis: The progression of cytological, genetic, and cellular changes that culminate in a malignant tumor.

Open pollination: Pollination by wind, insects, or other natural mechanisms.

Open reading frame: A long DNA sequence that is uninterrupted by a stop codon and encodes part or all of a protein. (

Operator: A prokaryotic regulatory element that interacts with a repressor to control the transcription of adjacent structural genes.

Orbital: In the approximation that each electron in a molecule has a distinct, independent wave function, the spatial distribution of an electron wave function corresponds to a molecular orbital. These, in turn, can be approximated as sums of contributions from the orbitals characteristic of the isolated atoms. An electron added to a molecule—or, similarly, one excited to a higher-energy state within a molecule—would occupy a state with a different wave function from the rest; an unoccupied state of this kind corresponds to an unoccupied molecular orbital. Orbital-symmetry effects on reaction rates arise when a reaction requires overlap between two lobes of the orbitals on each of two reagents:

if the algebraic signs of the wave functions in the facing lobes do not match, bond formation between those orbitals is prohibited.

Organelle: A cell structure that carries out a specialized function in the life of a cell.

Origin of replication: The nucleotide sequence at which DNA synthesis is initiated.

Overlapping reading frames: Start codons in different reading frames generate different polypeptides from the same DNA sequence.

Ovum: A female gamete.

Paleontology: The study of the fossil record of past geological periods and of the phylogenetic relationships between ancient and contemporary plant and animal species.

Palindromic sequence: A DNA locus whose 5'-to-3' sequence is identical on each DNA strand. The sequence is the same when one strand is read left to right and the other strand is read right to left. Recognition sites of many restriction enzymes are palindromic. See DNA.

pAMP: Ampicillin-resistant plasmid developed for this laboratory course.

Parasitism: The closee association of two or more dissimilar organisms where the association is harmful to at least one. See Commensalism, Parasitism, Symbiosis.

Particle bombardment: Using metal particles to blast DNA into cells for the purpose of genetic engineering. A gene gun is used to propel the metal particles.

Pathogen: Organism which can cause disease in another organism.

pBR322: A derivation of ColE1, one of the first plasmid vectors widely used.

PCR: Polymerase chain reaction. An amazingly sensitive technique that allows the specific amplification of extremely small amounts of particular DNA fragments using DNA polymerase and specific primers.

Pedigree: A diagram mapping the genetic history of a particular family.

Percent Accepted Mutation: A unit introduced by Dayhoff et al. to quantify the amount of evolutionary change in a protein sequence. 1.0 PAM unit, is the amount of evolution which will change, on average, 1% of amino acids in a protein sequence. A PAM(x) substitution matrix is a look-up table in which scores for each amino acid substitution have been calculated based on the frequency of that substitution in closely related proteins that have experienced a certain amount (x) of evolutionary divergence.

Persistence: Ability of an organism to remain in a particular setting for a period of time after it is introduced.

Pesticide: A substance that kills harmful organisms (for example, an insecticide or fungicide).

Phenotype: The observable characteristics of an organism, the expression of gene alleles (genotype) as an observable physical or biochemical trait. See Genotype.

Pheromone: A hormone-like substance that is secreted into the environment.

Phosphatase: An enzyme that hydrolyzes esters of phosphoric acid, removing a phosphate group.

Phosphodiester bond: A bond in which a phosphate group joins adjacent carbons through ester linkages. A condensation reaction between adjacent nucleotides results in a phosphodiester bond between 3' and 5' carbons in DNA and RNA.

Phospholipid: A class of lipid molecules in which a phosphate group is linked to glycerol and two fatty acyl groups. A chief component of biological membranes.

Phosphorylation: The addition of a phosphate group to a compound.

Physical map: A map showing physical locations on a DNA molecule, such as restriction sites, and sequence-tagged sites.

Piezoelectric material: A ferroelectric material in which an electrical potential difference is created due to mechanical deformation, or conversely, in which the application of a voltage causes dimensional changes in the material.

Piezoelectric: A piezoelectric material expands or contracts when a voltage is applied across it. Piezoelectric tubes are often used to move the tip in an AFM or STM. If a piezoelectric crystal is compressed, a voltage appears across it. This effect is sometimes used to measure displacements. A commonly used piezoelectric is PZT (Pb,Zr)TiO3.

Plaque: A clear spot on a lawn of bacteria or cultured cells where cells have been Iysed by viral infection.

Plasmid (p): A circular DNA molecule, capable of autonomous replication, which typically carries one or more genes encoding antibiotic resistance proteins. Plasmids can transfer genes between bacteria and are important tools of transformation for genetic engineers.

Pleiotrophy: The effect of a particular gene on several different traits.

Point mutation: A change in a single base pair of a DNA sequence in a gene.

Poly(A) polymerase: Catalyzes the addition of adenine

residues to the 3' end of pre-mRNAs to form the poly(A) tail.

Polyacrylamide gel electrophoresis: Electrophoresis through a matrix composed of a synthetic polymer, used to separate proteins, small DNA, or RNA molecules of up to 1000 nucleotides. Used in DNA sequencing.

Polyclonal antibodies: A mixture of immunoglobulin molecules secreted against a specific antigen, each recognizing a different epitope.

Polygenic: Controlled by or associated with more than one gene.

Polylinker: A short DNA sequence containing several restriction enzyme recognition sites that is contained in cloning vectors.

Polymer: A molecule composed of repeated subunits.

Polymerase (DNA): Synthesizes a double-stranded DNA molecule using a primer and DNA as a template.

Recessive gene: Characterized as having a phenotype expressed only when both copies of the gene are mutated or missing.

Recessive(-acting) oncogene, (anti-oncogene): A single copy of this gene is sufficient to suppress cell proliferation; the loss of both copies of the gene contributes to cancer formation.

Recessive: Refers to the member of a pair of alleles that fails to be expressed in the phenotype of the organism when the dominant member is present. Also refers to the phenotype of an individual that has only the recessive allele.

Recognition sequence (site): A nucleotide sequence—composed typically of 4, 6, or 8 nucleotides—that is recognized by a restriction endonuclease. Type II enzyrnes cut (and their corresponding modification enzymes methylate) within or very near the recognition sequence.

Recombinant DNA: The process of cutting and recombining DNA fragments from different sources as a means to isolate genes or to alter their structure and function.

Recombination frequency: The frequency at which crossing over occurs between two chromosomal loci—the probability that two loci will become unlinked during meiosis.

Refuge: A refuge is any host plant (non-Bt crop, potatoes, oats, sorghum, and some weeds) not producing Bt proteins or not being treated with conventional Bt formulations. The purpose of the refuge is to supply a source of susceptible insects that may mate with resistant insects of the same species emerging from nearby Bt crop fields. If the refuge is large enough, susceptible insect numbers should outnumber resistant individuals, and in theory, a greater number of the next generation offspring will be susceptible to the Bt crop.

Regulatory gene: A gene whose protein controls the activity of other genes or metabolic pathways.

Relaxed plasmid: A plasmid that replicates independently of the main bacterial chromosome and is present in 10-500 copies per cell.

Renature: The reannealing (hydrogen bonding) of single-stranded DNA and/or RNA to form a duplex molecule.

Replicon: A chromosomal region containing the DNA sequences necessary to initiate DNA replication processes.

Repressor: A DNA-binding protein in prokaryotes that blocks gene transcription by binding to the operator.

Resistance: The evolved ability of a species to survive during a stressful event or set of circumstances.

Restriction endonuclease (enzyme): A class of endonucleases that cleaves DNA after recognizing a specific sequence, such as BamH1 (GGATCC), EcoRI (GAATTC), and HindIII (AAGCTT). Type I. Cuts nonspecifically a distance greater than 1000 bp from its recognition sequence and contains both restriction and methylation activities. Type II. Cuts at or near a short, and often symmetrical, recognition sequence. A separate enzyme methylates the same recognition sequence. Type III. Cuts 24-26 bp downstream from a short, asymmetrical recognition sequence. Requires ATP and contains both restriction and methylation activities.

Restriction enzyme: A class of enzymes that cut DNA at specific sequences called restriction sites. Restriction enzymes were key to making genetic engineering possible. See RFLP.

Restriction-fragment-length polymorphism (RFLP): Differences in nucleotide sequence between alleles at a chromosomal locus result in restriction fragments of varying lengths detected by Southern analysis.

Retrovirus: A member of a class of RNA viruses that utilizes the enzyme reverse transcriptase to reverse copy its genome into a DNA intermediate, which integrates into the hostcell chromosome. Many naturally occurring cancers of vertebrate animals are caused by retroviruses.

Reverse genetics: Using linkage analysis and polymorphic markers to isolate a disease gene in the absence of a known metabolic defect, then using the DNA sequence of the cloned gene to predict the amino acid sequence of its encoded protein.

Reverse transcriptase (RNA-dependent DNA polymerase): An enzyme isolated from retrovirus-infected cells that synthesizes a complementary (c)DNA strand from an RNA template.

RFLP: Restriction fragment length polymorphism. RFLPs are generated by digesting DNA with restriction enzymes. The DNA is separated by size by gel electrophoresis. Slight differences in homologous fragments may exist between individuals. These length polymorphisms are RFLPs. See blot.

Rhizobia: Bacteria in a symbiotic relationship with leguminous plants that results in nitrogen fixation. See Nitrogen fixation.

Rhizosphere: The soils region on and around plant roots.

Ribosomal RNA (rRNA): The RNA component of the ribosome.

Ribosome: Cellular organelle that is the site of protein synthesis during translation. See Organelle, Translation.

Ribosome-binding site: The region of an mRNA molecule that binds the ribosome to initiate translation.

Risk assessment: Defined as a formalized basis for the objective evaluation of risk in a manner in which assumptions and uncertainties are clearly considered and presented.

RNA (ribonucleic acid): An organic acid composed of repeating nucleotide units of adenine, guanine, cytosine, and uracil, whose ribose components are linked by phosphodiester bonds.

RNA polymerase: Transcribes RNA from a DNA template.

RNA: Ribonucleic acid. The three basic types are messenger RNA (mRNA), ribosomal RNA (rRNA), and transfer RNA (tRNA).

Roundup Ready: A trademark for plants that are genetically engineered to be resistant to the herbicide Roundup (technical name: glyphosate).

Salmonella: A genus of rod-shaped, gram-negative bacteria that are a common cause of food poisoning.

Satellite RNA (viroids): A small, self-splicing RNA molecule that accompanies several plant viruses, including tobacco ringspot virus.

Scanning Near Field Optical Microscopy: A method for observing local optical properties of a surface that can be smaller than the wavelength of the light used. A type of scanning probe microscopy that maps the near-field optical properties of an interface.

Scanning Probe Microscopy (SPM): A method for imaging nanoscale features of surfaces by scanning a sensor (probe) over a surface. Near-field effects such as tunneling, van der Waals forces, local fields and more are serially detected at localized points on the surface and used to create an SPM image.

Scanning Thermal Microscopy (SThM): A type of scanning probe microscopy that maps the local temperature and thermal conductivity of an interface. A method for observing local temperatures and temperature gradients on a surface. [NTN]

Scanning Tunneling Microscope (STM): An instrument able to image conducting surfaces to atomic accuracy; has been used to pin: molecules to a surface. A device in which a sharp conductive tip is moved across a conductive surface close enough to permit a substantial tunneling current (typically a nanometer or less). In a common mode of operation, the voltage is kept constant and the current is monitored and kept constant by controlling the height of the tip above the surface; the result, under favorable conditions, is an atomic-resolution map of the surface reflecting a combination of topography and electronic properties. The STM has been used to manipulate atoms and molecules on surfaces. A type of scanning probe microscopy that maps the local electrical properties of an interface.

Schwann cell: Glial cell responsible for forming myelin sheaths in the peripheral nervous system.

Science: The process of developing a systematized knowledge of the world through the variation and testing of hypotheses. The pursuit of knowledge and understanding, from the Latin term *scientia*, which means 'knowledge'.

Selectable marker genes: Genetic engineers must be able to select cells that have received the transgene from those that have not. Selecting out transgenic cells is done by co-transforming the cells with the transgene plus an additional gene called a selectable marker gene. Selectable marker genes are genes that encode easily detectable traits making transgenic cells discernible from non-transgenic cells. The two most commonly used selectable marker genes encode the traits of herbicide and antibiotic resistance.

Selectable marker: A gene whose expression allows one to identify cells that have been transforrned or transfected with a vector containing the marker gene.

Self-pollination: Pollen of one plant is transferred to the female part of the same plant or another plant with the same genetic makeup.

Semiconservative replication: During DNA duplication, each strand of a parent DNA molecule is a template for the synthesis of its new complementary strand. Thus, one half of a preexisting DNA molecule is conserved during each round of replication.

Sequence hypothesis: Francis Crick's seminal concept that genetic information exists as a linear DNA code; DNA and protein sequence are colinear.

Sequence: Noun: the particular order of nucleotides in a DNA or RNA fragment. Verb: to determine the particular order of nucleotides in a strand of DNA.

Sequence-tagged site (STS): A unique (single-copy) DNA sequence used as a mapping landmark on a chromosome.

Sexual reproduction: The process where two cells (gametes) fuse to form one hybrid, fertilized cell. See Asexual reproduction, Gamete, Hybrid.

Signal transduction: The biochemical events that conduct the signal of a hormone or growth factor from the cell exterior, through the cell membrane, and into the cytoplasm. This involves a number of molecules, including receptors, pro- teins, and messengers.

Site-directed mutagenesis: The process of introducing spe- cific base-pair mutations into a gene.

Small nuclear RNA (snRNA): Short RNA transcripts of 100-300 bp that associate with proteins to form small nuclear ribonucleoprotein particles (snRNPs), which participate in RNA processing.

Somatic cell gene therapy: The repair or replacement of a defective gene within somatic tissue.

Somatic cell: Any nongerm cell that composes the body of an organism and which possesses a set of multiploid chromosomes (diploid in most organisms).

Southern hybridization (Southern blotting): A procedure in which DNA restriction fragments are transferred from an agarose gel to a nitrocellulose filter, where the denatured DNA is then hybridized to a radioactive probe (blotting).

Species: A classification of related organisms that can freely interbreed.

Terminator region: A DNA sequence that signals the end of transcription.

Tesla [T]: Unit of magnetic induction: 1T = 1 weber/m2 (also, 1T = 104 gauss).

Thermistor: A temperature-measuring device, that contains a resistor or semiconductor whose resistance varies with temperature.

Thermocouple: A temperature-measuring device, which contains a pair of end-joined dissimilar conductors in which an electromotive force is developed by thermoelectric effects when the joined ends and the free ends of the conductors are a different temperature.

Thiol: An SH group, or a molecule containing one. Also known as a sulfhydryl or mercapto group.

Tanscription: The copying of genetic information from DNA to messenger RNA (mRNA).

Transduction: The transfer of DNA sequences from one bacterium to another via lysogenic infection by a bacteriophage (transducing phage).

Transfection: The uptake and expression of a foreign DNA sequence by cultured eukaryotic cells.

Transformant: In prokaryotes, a cell that has been genetically altered through the uptake of foreign DNA. In higher eukaryotes, a cultured cell that has acquired a malignant phenotype.

Transformation: (1) Transformation is the step in the genetic engineering process where a new gene (transgene) is delivered into the nucleus of a plant cell and inserts into a chromosome where it is passed on to progeny. Transformation means to genetically change a living thing. (2) In prokaryotes, the natural or induced uptake and expression of a foreign DNA sequence—typically a recombinant plasmid in experimental systems. In higher eukaryotes, the conversion of cultured cells to a malignant phenotype—typically through infection by a tumor virus or transfection with an oncogene.

Transformation efficiency: The number of bacterial cells that uptake and express plasmid DNA divided by the mass of plasmid used (in transformants/microgram).

Transforming oncogene: A gene that upon transfection converts a previously immortalized cell to the malignant phenotype.

Transgene: A foreign gene incorporated by transformation.

Transgenic: A genetically engineered plant created through transformation is sometimes referred to as a transgenic.

Transgenic animal: Genetically enginnered animal or offspring of genetically engineered animals. The transgenic animal usually contains material from at lease one unrelated organism, such as from a virus, plant, or other animal. See Transgenic.

Trypsin: A proteolytic enzyme that hydrolyzes peptide bonds on the carboxyl side of the amino acids arginine and lysine.

TSCA: The Toxic Substances Control Act. See Environmental Protection Agency.

Vaccine: A preparation of dead or weakened pathogen, or of derived antigenic determinants, that is used to induce formation of antibodies or immunity against the pathogen.

Vaccinia: The cowpox virus used to vaccinate against smallpox and, experimentally, as a carrier of genes for antigenic determinants cloned from other disease organisms.

Variable surface glycoprotein (VSG): One of a battery of antigenic determinants expressed by a microorganism to elude immune detection.

Variation: Differences in the frequency of genes and traits among individual organisms within a population.

Vasculocyte: In medical nanorobotics, a theorized (nanorobotic) device capable of performing repairs of an injured vascular luminal surface.

Vasculography: A physical description (and mapping) of the human vascular system.

Vector: An autonomously replicating DNA molecule into which foreign DNA fragments are inserted and then propagated in a host cell. Also living carriers of genetic material (such as pollen) from plant to plant, such as insects.

Vector: In cell biology, an agent (virus or plasmid) used to transmit genetic material to a cell or organism.

Vegetal pole: The end at which most of the yolk is located in an animal egg. The end opposite the animal pole.

Viral oncogene: A viral gene that contributes to malignancies in vertebrate hosts.

1

Biotech Information & Database: An Introductory Overview

Over the past few decades, major advances in the field of molecular biology, coupled with advances in genomic technologies, have led to an explosive growth in the biological information generated by the scientific community. This deluge of genomic information has, in turn, led to an absolute requirement for computerized databases to store, organize, and index the data and for specialized tools to view and analyze the data.

A biological database is a large, organized body of persistent data, usually associated with computerized software designed to update, query, and retrieve components of the data stored within the system. A simple database might be a single file containing many records, each of which includes the same set of information. For example, a record associated with a nucleotide sequence database typically contains information such as contact name, the input sequence with a description of the type of molecule, the scientific name of the source organism from which it was isolated, and often, literature citations associated with the sequence.

For researchers to benefit from the data stored in a database, two additional requirements must be met:

- easy access to the information
- a method for extracting only that information needed to answer a specific biological question

At NCBI, many of our databases are linked through a unique search and retrieval system, called Entrez. Entrez (pronounced ahn' tray) allows a user to not only access and retrieve specific information from a single database but to access integrated information from many NCBI databases. For example, the Entrez Protein database is cross-linked to the Entrez Taxonomy database. This allows a researcher to find taxonomic information (taxonomy is a division of the natural sciences that deals with the classification of animals and plants) for the species from which a protein sequence was derived.

> Biology in the 21st century is being transformed from a purely lab-based science to an information science as well.

Bioinformatics is the field of science in which biology, computer science, and information technology merge to form a single discipline. The ultimate goal of the field is to enable the discovery of new biological insights as well as to create a global perspective from which unifying principles in biology can be discerned. At the beginning of the "genomic revolution", a bioinformatics concern was the creation and maintenance of a database to store biological information, such as nucleotide and amino acid sequences. Development of this type of database involved not only design issues but the development of complex interfaces whereby researchers could both access existing data as well as submit new or revised data.

Ultimately, however, all of this information must be combined to form a comprehensive picture of normal cellular activities so that researchers may study how these activities are altered in different disease states. Therefore, the field of bioinformatics has evolved such that the most pressing task now involves the analysis and interpretation of various types of data, including nucleotide and amino acid sequences, protein domains, and protein structures. The actual process of analyzing and interpreting data is referred to as computational biology. Important sub-disciplines within bioinformatics and computational biology include:

- the development and implementation of tools that enable efficient access to, and use and management of, various types of information
- the development of new algorithms (mathematical formulas) and statistics with which to assess relationships among members of large data sets, such as methods to locate a gene within a sequence, predict protein structure and/or function, and cluster protein sequences into families of related sequences

The rationale for applying computational approaches to facilitate the understanding of various biological processes includes:

- a more global perspective in experimental design
- the ability to capitalize on the emerging technology of database-mining - the process by which testable hypotheses are generated regarding the function or structure of a gene or protein of interest by identifying similar sequences in better characterized organisms.

Although a human disease may not be found in exactly the same form in animals, there may be sufficient data for an animal model that allow researchers to make inferences about the process in humans.

Evolutionary Biology

New insight into the molecular basis of a disease may

come from investigating the function of homologs of a disease gene in model organisms. In this case, homology refers to two genes sharing a common evolutionary history. Scientists also use the term homology, or homologous, to simply mean similar, regardless of the evolutionary relationship.

> NCBI's COGs database has been designed to simplify evolutionary studies of complete genomes and to improve functional assignment of individual proteins.

Equally exciting is the potential for uncovering evolutionary relationships and patterns between different forms of life. With the aid of nucleotide and protein sequences, it should be possible to find the ancestral ties between different organisms. Thus far, experience has taught us that closely related organisms have similar sequences and that more distantly related organisms have more dissimilar sequences. Proteins that show a significant sequence conservation, indicating a clear evolutionary relationship, are said to be from the same protein family. By studying protein folds (distinct protein building blocks) and families, scientists are able to reconstruct the evolutionary relationship between two species and to estimate the time of divergence between two organisms since they last shared a common ancestor.

> Phylogenetics is the field of biology that deals with identifying and understanding the relationships between the different kinds of life on earth.

Protein Modeling

The process of evolution has resulted in the production of DNA sequences that encode proteins with specific functions. In the absence of a protein structure that has been determined by X-ray crystallography or nuclear magnetic resonance (NMR) spectroscopy, researchers can try to predict the three-dimensional structure using protein or molecular modeling.

This method uses experimentally determined protein structures (templates) to predict the structure of another protein that has a similar amino acid sequence (target).

Although molecular modeling may not be as accurate at determining a protein's structure as experimental methods, it is still extremely helpful in proposing and testing various biological hypotheses. Molecular modeling also provides a starting point for researchers wishing to confirm a structure through X-ray crystallography and NMR spectroscopy. Because the different genome projects are producing more sequences and because novel protein folds and families are being determined, protein modeling will become an increasingly important tocl for scientists working to understand normal and disease-related processes in living organisms.

The Four Steps of Protein Modeling

- Identify the proteins with known three-dimensional structures that are related to the target sequence
- Align the related three-dimensional structures with the target sequence and determine those structures that will be used as templates
- Construct a model for the target sequence based on its alignment with the template structure(s)
- Evaluate the model against a variety of criteria to determine if it is satisfactory

Genome Mapping

Genomic maps serve as a scaffold for orienting sequence information. A few years ago, a researcher wanting to localize a gene, or nucleotide sequence, was forced to manually map the genomic region of interest, a time-consuming and often painstaking process. Today, thanks to new technologies and the influx of sequence data, a number of high-quality, genome-wide maps are available to the scientific community for use in their research.

Computerized maps make gene hunting faster, cheaper, and more practical for almost any scientist. In a nutshell, scientists would first use a genetic map to assign a gene to a relatively small area of a chromosome. They would then use a physical map to examine the region of interest close up, to determine a gene's precise location. In light of these advances, a researcher's burden has shifted from mapping a genome or genomic region of interest to navigating a vast number of Web sites and databases.

Map Viewer: A Tool for Visualizing Whole Genomes or Single Chromosomes

NCBI's Map Viewer is a tool that allows a user to view an organism's complete genome, integrated maps for each chromosome (when available), and/or sequence data for a genomic region of interest. When using Map Viewer, a researcher has the option of selecting either a "Whole-Genome View" or a "Chromosome or Map View". The Genome View displays a schematic for all of an organism's chromosomes, whereas the Map View shows one or more detailed maps for a single chromosome. If more than one map exists for a chromosome, Map Viewer allows a display of these maps simultaneously.

Using Map Viewer, researchers can find answers to questions such as:

- Where does a particular gene exist within an organism's genome?
- Which genes are located on a particular chromosome and in what order?
- What is the corresponding sequence data for a gene that exists in a particular chromosomal region?
- What is the distance between two genes?

The rapidly emerging field of bioinformatics promises to lead to advances in understanding basic biological processes and, in turn, advances in the diagnosis, treatment, and prevention of many genetic diseases. Bioinformatics has transformed the discipline of biology from a purely lab-based science to an information science as well. Increasingly, biological studies begin with a scientist conducting vast numbers of database and Web site searches to formulate specific hypotheses or to design large-scale experiments. The implications behind this change, for both science and medicine, are staggering.

Computational Chemistry

Reviews in Computational Chemistry by Lipkowitz and Boyd is an excellent series of books which does just that. Unlike most review series, it includes introductory and tutorial as well as more advanced material.

The World-Wide Web Virtual Library: Chemistry is an excellent starting place to look for any chemistry related information.

The Computational Chemistry Archive is more specifically focused on issues surrounding computational chemistry.

Infochem has many links to Molecular Modelling Software.

Chemistry Software and Information Resources of the National HPCC.

QCPE (The Quantum Chemistry Program Exchange) provides a wide range of computational chemistry software, often free.

NetSci's list of molecular modeling software and their computational chemistry page

A list of chemistry resources on the internet is maintained at Rensselaer.

NIH has a page on molecular modeling and has a guide to software.

An online-text which discusses aspects of quantum mechanics, the Born-Oppenheimer approximation, and molecular mechanics.

Molecular Modeling

One may have some direct experience with a few software packages:

MSI's Cerius 2 provides an excellent but relatively expensive package which runs on several commercial platforms (SGI, RS6000, and the like).

Hypercube's Hyperchem under Windows on the PC provides excellent capabilities at the price.

CambridgeSoft provides a package that runs on the Mac and under Windows. It is more limited in its modeling capabilities: it can effectively do energy minimizations on structures with perhaps a hundred atoms (their minimizer does only steepest descent). They provide truly excellent facilities for making black & white pictures of molecules, and equally excellent facilities for drawing journal-quality 2-dimensional chemical structures using standard notation.

Gaussian is an ab initio program that usually costs over $10,000 but which is now available on the PC for under $1,000. While it has an antediluvian user interface (designed for punch cards) it is the de facto standard against which competing companies (like the startup Schrodinger) compare themselves and is well known in the computational chemistry literature. They can be reached by e-mail at info@gaussian.com.

While I haven't had direct experience with it, Jay Ponder (who gave an excellent talk on molecular mechanics at the

First Foresight conference on Molecular Nanotechnology) has made available a molecular mechanics and dynamics package written in Fortran 77 which is free (including source code). To quote their description:

TINKER is a complete package for performing molecular mechanics calculations. It is intended to serve as a platform for algorithm development and parameterization, while still being efficient enough for most production work. The currently available potentials include MM2 and MM3, AMBER-95, AMBER/OPLS, CHARMM22 and our own TINKER potential. Many other literature potentials can be implemented via external parameter files; parameter sets for ENCAD and MMFF-94 are under development.

Programs are provided to perform many functions including the following: (1) energy minimization over Cartesian or torsional coordinates via conjugate gradient, variable metric or truncated Newton methods, (2) molecular dynamics with periodic boundaries and options to control temperature and pressure, (3) normal mode vibrational analysis, (4) distance geometry including a new, efficient random atom-pair metrization, (5) simulated annealing with a choice of cooling protocols, (6) analysis and breakdown of single point potential energies, (7) verification of analytical derivatives of the standard potential and/or user defined functions, (8) location of a transition state between two minima, (9) free energy perturbation calculations, (10) fitting of intermolecular potential parameters to crystal structure and thermodynamic data, and (11) diffusion equation method and Gaussian density annealing approaches to optimization via smoothed potential energy surfaces.

Major changes in the field of biology have taken place over the last few decades. Advances in biotechnology and accomplishments such as the mapping of the entire human genome and the genomes of many other organisms are producing a better understanding of biological systems and living organisms. Large amounts of biological information and data have been and continue to be generated through this research. Bioinformatics arose from the need to organize, store, and analyse this information.

Bioinformatics is the use of computer software and programs to organize, store and analyse biological information and data to better understand biological systems. It is the field of science where biology, computer science and information technology converge as a single discipline. Researchers use advanced computer and statistical techniques to wade through and analyse large amounts of biological data created through modern biotechnologies referred to as the "omics" technologies – such as genomics and proteomics. These "omics" technologies focus on different aspects of biological systems. For example:

- Genomics is the analysis of the entire genetic make-up of an organism. It includes a study of the functions of genes in cells, organs and organisms.
- Proteomics focusses on the total protein make-up of a cell at any given time.

These "omics" technologies generate massive amounts of biological data through which it would be impossible to navigate without the use of computer systems. The data includes sequences of amino acids and nucleotides that underlie genes and proteins. The goal of bioinformatics is to translate the complex data gathered into usable knowledge. Bioinformatics typically includes three main areas:

- Developing new statistics and data to be studied and used to assess relationships among the different sets of biological data.

- Analysing and interpreting the various types of data, which include nucleotide and amino acid sequences and the properties of various proteins.
- Developing and implementing tools to enable efficient access and management of different types of pertinent information.

The field of bioinformatics emerged in the early 1980s with the creation of the GenBank database. GenBank, started by the US Department of Energy, stores DNA sequence information obtained from a wide range of organisms. In the early days, GenBank was a small scale operation with a roomful of technicians sitting at keyboards entering the DNA sequence information published in academic journals.

The advent of the Internet allowed researchers to access data in GenBank from all over the world for free. The DNA sequence data in GenBank grew rapidly with the emergence of highly sophisticated gene sequencing tools. Private companies joining the sequencing race with parallel projects created huge databases of their own.

Several services emerged as the need for access to bioinformatics increased. The two most significant were the European Molecular Biology Network (EMBnet) and the United States National Center for Biotechnology Information (NCBIO). In Canada, the Canada Institute for Scientific and Technical Information (CISTI) of the National Research Council of Canada (NRC) operated a bioinformatics server, Molecular Biology Database Service (MBDS), from 1987 to 1996. The Canadian Bioinformatics Resource (CBR) was created in 1995 to replace MBDS in providing bioinformatics resources and services to Canadian researchers.

Bioinformatics Applications

Most bioinformatics tools currently available deal with the structural and functional aspects of genes and proteins – looking at what genes and proteins look like, where they are located and what they do. Applications of bioinformatics allow researchers to accomplish the following:

- Compare mapped genomes of any species to identify similarities and differences among organisms.
- Researchers can access genetic information from around the world. Various searchable databases, many of which are on the Internet, are used to identify DNA sequences of identified genes.
- Databases also allow quick access to information about proteins. Researchers retrieve a sequence using a protein name, or retrieve a protein using the amino acid sequence.
- Sequence translation programs enable users to analyse nucleotide sequences. The programs convert the DNA sequences into protein sequences or protein sequences into their complementary DNA (cDNA).
- Sequencing tools are used by bioinformatists to find out more about the characteristics of genetic material, including proteins. Once the protein sequences are analysed, the researchers can use software programs to search databases to find similar sequences. Other programs allow researchers to look at the three-dimensional shape of the proteins and nucleotides. Since shape is a critical determinant of function, these programs allow researchers to gain clues for identifying the roles of the sequences.

Examples of Common Bioinformatics Tools

Data produced through research from all over the world

is collected and organized in databases specialized in individual subjects. Computational tools are used to efficiently analyse this data. Many areas of bioinformatics focus on predicting the biological functions of genes, proteins, and their parts based on the structural data compiled. Examples of commonly used tools include:

- BLAST (Basic Local Alignment Search Tool) — This tool allows researchers to identify similarities between their own nucleotide or protein sequence with those in the public databases. It is used to identify whether a given sequence is new, similar to a known sequence (e.g. other proteins), or contains specific protein characteristics which may give a clue to the possible role of the sequence.
- OMIM (Online Mendelian Inheritance in Man) — This database contains information about human genes and genetic disorders already mapped in the human genome. Typically, the information contained includes a summary of each disease, including symptoms, genetic mapping information, related references, and a direct link to the DNA sequence in GenBank if the gene responsible for the disease has been cloned. It also contains mapping information on genes not yet linked to diseases.

GeneMap98 — This database lists genes mapped in the human genome. It is helpful to researchers who are looking to identify genes for diseases with simple traits or symptoms since the map can display genes within a location defined by the user.

Bioinformatics is an interdisciplinary field that addresses biological problems using computational techniques. The field is also often referred to as computational biology. It plays a key role in various areas like functional genomics, structural genomics, and proteomics amongst others, and forms a key component in biotechnology and pharmaceutical sector. Bioinformatics and computational biology involve the use of techniques from applied mathematics, informatics, statistics, and computer science to solve biological problems. Research in computational biology often overlaps with systems biology. Major research efforts in the field include sequence alignment, gene finding, genome assembly, protein structure alignment, protein structure prediction, prediction of gene expression and protein-protein interactions, and the modeling of evolution. The terms *bioinformatics* and *computational biology* are often used interchangeably, although the former typically focuses on algorithm development and specific computational methods, while the latter focuses more on hypothesis testing and discovery in the biological domain. Although this distinction is used by National Institutes of Health in their working definitions of Bioinformatics and Computational Biology, it is clear that there is a tight coupling of developments and knowledge between the more hypothesis-driven research in computational biology and technique-driven research in bioinformatics. Computational biology also includes lesser-known but equally important sub-disciplines such as computational biochemistry and computational biophysics. A common thread in projects in bioinformatics and computational biology is the use of mathematical tools to extract useful information from noisy data produced by high-throughput biological techniques such as genomics (The field of data mining overlaps with computational biology in this regard). A representative problem in bioinformatics is the assembly of high-quality DNA sequences from fragmentary "shotgun" DNA sequencing, while in computational biology, a representative problem might be statistical testing of a hypothesis of common gene regulation using data from mRNA microarrays or mass spectrometry.

Sequence analysis

Since the Phage Ö-X174 was sequenced in 1977, the DNA sequences of more and more organisms have been decoded and stored in electronic databases. This data is analyzed to determine genes that code for proteins, as well as regulatory sequences. A comparison of genes within a species or between different species can show similarities between protein functions, or relations between species (the use of molecular systematics to construct phylogenetic trees). With the growing amount of data, it long ago became impractical to analyze DNA sequences manually. Today, computer programs are used to search the genome of thousands of organisms, containing billions of nucleotides. These programs can compensate for mutations (exchanged, deleted or inserted bases) in the DNA sequence, in order to identify sequences that are related, but not identical. A variant of this sequence alignment is used in the sequencing process itself. The so-called shotgun sequencing technique (which was used, for example, by The Institute for Genomic Research to sequence the first bacterial genome, *Haemophilus influenza*) does not give a sequential list of nucleotides, but instead the sequences of thousands of small DNA fragments (each about 600-800 nucleotides long). The ends of these fragments overlap and, when aligned in the right way, make up the complete genome. Shotgun sequencing yields sequence data quickly, but the task of assembling the fragments can be quite complicated for larger genomes. In the case of the Human Genome Project, it took several months of CPU time (on a circa-2000 vintage DEC Alpha computer) to assemble the fragments. Shotgun sequencing is the method of choice for virtually all genomes sequenced today, and genome assembly algorithms are a critical area of bioinformatics research. Another aspect of bioinformatics in sequence analysis is the automatic search for genes and regulatory sequences within a genome. Not all of the nucleotides within a genome are genes. Within the genome of higher organisms, large parts of the DNA do not serve any obvious purpose. This so-called junk DNA may, however, contain unrecognized functional elements. Bioinformatics helps to bridge the gap between genome and proteome projects, for example in the use of DNA sequence for protein identification.

Genome annotation

In the context of genomics, annotation is the process of marking the genes and other biological features in a DNA sequence. The first genome annotation software system was designed in 1995 by Owen White, who was part of the team that sequenced and analyzed the first genome of a free-living organism to be decoded, the bacterium Haemophilus influenzae. Dr. White built a software system to find the genes (places in the DNA sequence that encode a protein), the transfer RNA, and other features, and to make initial assignments of function to those genes. Most current genome annotation systems work similarly, but the programs available for analysis of genomic DNA are constantly changing and improving.

Computational evolutionary biology

Evolutionary biology is the study of the origin and descent of species, as well as their change over time. Informatics has assisted evolutionary biologists in several key ways; it has enabled researchers to:

- trace the evolution of a large number of organisms by measuring changes in their DNA, rather than through physical taxonomy or physiological observations alone,
- more recently, compare entire genomes, which permits the study of more complex evolutionary

events, such as gene duplication, lateral gene transfer, and the prediction of bacterial speciation factors,

- build complex computational models of populations to predict the outcome of the system over time
- track and share information on an increasingly large number of species and organisms

Future work endeavors to reconstruct the now more complex tree of life. The area of research within computer science that uses genetic algorithms is sometimes confused with computational evolutionary biology. Work in this area involves using specialized computer software to improve equations, algorithms, or integrated circuit designs. It is inspired by evolutionary principles such as replication, diversification through recombination or mutation, fitness, survival through selection or culling, and iteration, collectively called a Darwinian machine or Darwinian ratchet.

Measuring biodiversity

Biodiversity of an ecosystem might be defined as the total genomic complement of a particular environment, from all of the species present, whether it is a biofilm in an abandoned mine, a drop of sea water, a scoop of soil, or the entire biosphere of the planet Earth. Databases are used to collect the species names, descriptions, distributions, genetic information, status and size of populations, habitat needs, and how each organism interacts with other species. Specialized software programs are used to find, visualize, and analyze the information, and most importantly, communicate it to other people. Computer simulations model such things as population dynamics, or calculate the cumulative genetic health of a breeding pool (in agriculture) or endangered population (in conservation). One very exciting potential of this field is that entire DNA sequences, or genomes of endangered species can be preserved, allowing the results of Nature's genetic experiment to be remembered *in silico*, and possibly reused in the future, even if that species is eventually lost.

Gene expression analysis

The expression of many genes can be determined by measuring mRNA levels with multiple techniques including microarrays, expressed cDNA sequence tag (EST) sequencing, serial analysis of gene expression (SAGE) tag sequencing, massively parallel signature sequencing (MPSS), or various applications of multiplexed in-situ hybridization. All of these techniques are extremely noise-prone and/or subject to bias in the biological measurement, and a major research area in computational biology involves developing statistical tools to separate signal from noise in high-throughput gene expression studies. Such studies are often used to determine the genes implicated in a disorder: one might compare microarray data from cancerous epithelial cells to data from non-cancerous cells to determine the transcripts that are up-regulated and down-regulated in a particular population of cancer cells.

Regulation analysis

Regulation is the complex orchestra of events starting with an extra-cellular signal and ultimately leading to the increase or decrease in the activity of one or more protein molecules. Bioinformatics techniques have been applied to explore various steps in this process. For example, promoter analysis involves the elucidation and study of sequence motifs in the genomic region surround the coding region of a gene. These motifs influence the extent to which that region is transcribed into mRNA. Expression data can be used to infer gene regulation: one might compare microarray data from a wide variety of states of an organism to form hypotheses about the genes involved in each state. In a single-cell organism, one might compare stages of the cell cycle, along with various stress conditions (heat shock, starvation, etc.). One can then apply clustering algorithms to that expression data to determine which genes are co-expressed. Further analysis could take a variety of directions: one 2004 study analyzed the promoter sequences of co-expressed (clustered together) genes to find common regulatory elements and used machine learning techniques to identify the promoter elements involved in regulating each cluster.

Protein expression analysis

Protein microarrays and high throughput (HT) mass spectrometry (MS) can provide a snapshot of the proteins present in a biological sample. Bioinformatics is very much involved in making sense of protein microarray and HT MS data; the former approach faces similar problems as with microarrays targeted at mRNA, the latter involves the problem of matching large amounts of mass data against predicted masses from protein sequence databases, and the complicated statistical analysis of samples where multiple, but incomplete peptides from each protein are detected.

Analysis of mutations in cancer

Massive sequencing efforts are currently underway to identify point mutations in a variety of genes in cancer. The sheer volume of data produced requires automated systems to read sequence data, and to compare the sequencing results to the known sequence of the human genome, including known germline polymorphisms. Oligonucleotide microarrays, including comparative genomic hybridization and single nucleotide polymorphism arrays, able to probe simultaneously up to several hundred thousand sites throughout the genome are being used to identify chromosomal gains and losses in cancer. Hidden Markov model and change-point analysis methods are being developed to infer real copy number changes from often-noisy data. Further informatics approaches are being developed to understand the implications of lesions found to be recurrent across many tumors. Some modern tools (e.g. Quantum 3.1) provide tool for changing the protein sequence at specific sites through alterations to its amino acids and predict changes in the bioactivity after mutations.

Structure prediction

Protein structure prediction is another important application of bioinformatics. The amino acid sequence of a protein, the so-called *primary structure*, can be easily determined from the sequence on the gene that codes for it. In the vast majority of cases, these primary structures uniquely determine a structure in its native environment. (Of course, there are exceptions, such as the bovine spongiform encephalopathy - aka Mad Cow Disease - prion.) Knowledge of this structure is vital in understanding the function of the protein. For lack of better terms, structural information is usually classified as one of *secondary*, *tertiary* and *quaternary* structures. A viable general solution to such predictions remains an open problem. As of now, most efforts have been directed towards heuristics that work most of the time. One of the key ideas in bioinformatics research is the notion of homology. In the genomic branch of bioinformatics, homology is used to predict the function of a gene: if the sequence of gene *A*, whose function is known, is homologous to the sequence of gene *B,* whose function is unknown, one could infer that B may share A's function. In the structural branch of bioinformatics homology is used to determine which parts of the protein are important in structure formation and interaction with other proteins. In a technique called homology modeling, this information is used to predict the structure of a protein once the structure of a homologous protein is known. This currently remains the only way to predict protein structures reliably. One example of this is the similar protein homology between hemoglobin in humans and the hemoglobin in legumes (leg hemoglobin). Both serve the same purpose of transporting oxygen in both organisms. Though both of these proteins have completely different amino acid sequences, their protein structures are virtually identical, which reflects their near identical purposes. Other techniques for predicting protein structure include protein threading and *de novo* (from scratch) physics-based modeling.

Comparative genomics

The core of comparative genome analysis is the establishment of the correspondence between genes (orthology analysis) or other genomic features in different organisms. It is these inter-genomic maps that make it possible to trace the evolutionary processes responsible for the divergence of two genomes. A multitude of evolutionary events acting at various organizational levels shape genome evolution. At the lowest level, point mutations affect individual nucleotides. At a higher level, large chromosomal segments undergo duplication, lateral transfer, inversion, transposition, deletion and insertion. Ultimately, whole genomes are involved in processes of hybridization, polyploidization and endosymbiosis, often leading to rapid speciation. The complexity of genome evolution poses many exciting challenges to developers of mathematical models and algorithms, who have recourse to a spectra of algorithmic, statistical and mathematical techniques, ranging from exact, heuristics, fixed parameter and approximation algorithms for problems based on parsimony models to Markov Chain Monte Carlo algorithms for Bayesian analysis of problems based on probabilistic models. Many of these studies are based on the homology detection and protein families' computation.

Modeling biological systems

Systems biology involves the use of computer simulations of cellular subsystems (such as the networks of metabolites and enzymes which comprise metabolism, signal transduction pathways and gene regulatory networks) to both analyze and visualize the complex connections of these cellular processes. Artificial life or virtual evolution attempts to understand evolutionary processes via the computer simulation of simple (artificial) life forms.

High-throughput image analysis

Computational technologies are also used to accelerate or fully automate the processing, quantification and analysis of large amounts of high-information-content Biomedical imagery. Modern image analysis systems augment the observers' ability to make measurements from a large or complex set of images, by improving accuracy, objectivity, or speed. A fully developed analysis system may completely replace the observer. While these systems are not unique to biology related imagery, their application to biologic problems continue to provide unique challenges and solutions, placing several imagery application under the umbrella of Bioinformatics. These systems are in the process of becoming more important for both diagnostics and research. Some examples:

- high-throughput and high-fidelity quantification and sub-cellular localization (high-content screening, cytohistopathology)
- morphometrics are used to analyze pictures of embryos to track and to predict the fate of cell clusters during morphogenesis

- clinical image analysis and visualization
- determine the real-time air-flow patterns in breathing lungs of living individuals before and during challenge
- quantify occlusion size in real-time imagery from the development of and recovery during arterial injury
- making behavioral observations from extended video recordings of laboratory animals
- infrared measurements for metabolic activity determination

Software tools

The computational biology tool best-known among biologists is probably BLAST, an algorithm for searching large sequence (protein, DNA) databases. NCBI provides a popular implementation that searches their massive sequence databases. Bioinformatic meta search engines (Entrez, Bioinformatic Harvester) help finding relevant information from several databases. There are also free Web-based software designed for structural bioinformatics such as STING. Computer scripting languages such as Perl and Python are often used to interface with biological databases and parse output from bioinformatics programs. Communities of bioinformatics programmers have set up free/open source projects such as EMBOSS, Bioconductor, BioPerl, BioLinux, BioPython, BioRuby, and BioJava which develop and distribute shared programming tools and objects (as program modules) that make bioinformatics easier. An integrated software workbench consisting of many free/open source tools described above and many others is known as VigyaanCD. Taverna an open-source bioinformatics workbench that utilizes a workflow model of experimental design. Taverna is included as part of the myGRID package of e-science software. Quantum 3.1 is an example of the bioinformatics post-QSAR technology applying quantum and molecular physics instead of statistical methods. Genevestigator is an example of how large-scale gene expression microarray data is used to predict gene function based on contextual information. More recently, SOAP-based interfaces have been developed for a wide variety of bioinformatics applications such as blast, fasta, EMBOSS, clustalw, t-coffee, MUSCLE and many others. These are available from the EBI at EBI Web Services.

REFERENCES

Abraham, E. J., et al., "Insulinotropic hormone glucagons-like peptide-1 differentiation of human pancreatic islet-derived progenitor cells into insulinproducing cells," Endocrinology 143: 3152-3161 (2002).

Assady, S., et al., "*Insulin* production by human embryonic stem cells," Diabetes 50: 1691-1697 (2001).

Atkinson, M. A. and Leiter, E. H., "The NOD mouse model of type 1 diabetes: As good as it gets?" Nature Medicine 5: 601-604 (1999).

Baker, K. S., et *al.*, "*New* Malignancies After Blood or Marrow Stem-Cell Transplantation in Children and Adults: Incidence and Risk Factors," Journal of Clinical Investigation 21: 1352-1358 (2003).

Brivanlou, A. H., et *al.*, "Stem cells. Setting standards for human embryonic stem cells," Science 300: 913-916 (2003).

Chen, Y., *et al.*, "*Embryonic* stem cells generated by nuclear transfer of human somatic nuclei into rabbit oocytes," Cell Research 13: 251-264 (2003).

Daly, G., "Cloning and Stem Cells—Handicapping the Political and Scientific Debates," New England Journal of Medicine 349: 211-212 (2003).

Drukker, M., et al., "*Charac*terization of the expression of MHC proteins in human embryonic stem cells," Proceedings of the National Academy of Sciences of th*e United States of America 99: 9864-9869 (2002).*

Faraci, M., et *al., "Severe ne*urologic complications after hematopoietic stem cell transplantation in children," Neurology 59: 1895-1904 (2002).

Hogan, P., et al., "Economic Costs of Diabetes in the US in 2002," Diabetes Care 26: 917-932 (2003).

Majumdar, M. K., et al., "Characterization and functionality of cell surface molecules on human mesenchymal stem cells," Journal of Biomedical Science 10: 228-241 (2003).

Maxam, A. M. & Gilbert,W. A new method for sequencing DNA.

McCarthy PM, Rose EA, Macmanus Q, et al. Clinical experience with the Novacor ventricular assist system. J Thorac Cardiovasc Surg. 1991; 102:578-587.

McEwan PE, Gray GA, Sherry L, Webb DJ, Kenyon CJ. Differential effects of angiotensin II on cardiac cell proliferation and intramyocardial perivascular fibrosis in vivo. Circulation 1998;98:2765-2773.

Mendel, G.Versuche über Pflanzen-Hybriden.*Verhandlungen des naturforschenden Vereines, Abhandlungen, Brünn* **4**, 3–47 (1866).

Migeon, B.R., Chowdry, A.K., Dunston, J.A., and McIntosh, I. 2001. Identification of *TSIX*, encoding an RNA antisense to human *XIST*, reveals differences from its murine counterpart: Implications for X inactivation. Am. J. Human Genet. 69, 951-960.

National Center for Health Statistics, "Births: Final Data for 2001." National Vital Statistics Reports 51(2) (2002), available at http://www.cdc.gov/nchs/data/nvsr/ nvsr51/nvsr51_02.pdf.

National Institutes of Health Revitalization Act of 1993, Pub. L. No. 103-43, § 121(c), 107 Stat. 122 (1993), repealing 45 C.F.R. § 46.204(d).

National Institutes of Health, *Report of the Human Embryo Research Panel*, Bethesda, MD: NIH (1994).

National Research Council.*Mapping and Sequencing the Human Genome* (National Academy Press,Washington DC, 1988).

Nelissen-Vrancken H, Debets J, Snoeckx L, Daemen M, Smits J. Time-related normalization of maximal coronary flow in isolated perfused hearts of rats with myocardial infarction. Circulation 1996;93:349-355.

Nester, E.W., Gordon, M.D., Amasino, R.M., Yanofsky, M.F. 1984. Crown gall: a molecular and physiological analysis. Annual Review of Plant Pathology 35:387-413.

Nidorf SM, Siu SC, Galambos G, Weyman AE, Picard MH. Benefit of late coronary reperfusion on ventricular morphology and function after myocardial infarction. J Am Coll Cardiol 1992;20:307-313.

Nirenberg, M.W. The genetic code: II. *Sci.Am.* **208,** 80–94 (1963).

Norwitz, E. R., et al., "Implantation and the survival of early pregnancy," *New England Journal of Medicine* 345(19): 1400-1408 (2001)

Ogawa M, Kizumoto M, Nishikawa S, Fujimoto T, Kodama H, Nishikawa SI. Blood 1999;93:1168-1177.

Okarma, T., Presentation at the September 4, 2003, meeting of the President's Council on Bioethics, Washington, D.C., available at www.bioethics.gov.

Onyango, P., Jiang, S., Uejima, H., Shamblott, M.J., Gearhart, J.D., Cui, H., and Feinberg, A.P. 2002. Monoallelic expression and methylation of imprinted genes in human and mouse embryonic germ cell lineages. Proc. Natl. Acad. Sci. USA 99, 10599-10604.

Onyango, P., Jiang, S., Uejima, H., Shamblott, M.J., Gearhart, J.D., Cui, H., and Feinberg, A.P. 2002. Monoallelic expression and methylation of imprinted genes in human and mouse embryonic germ cell lineages. Proc. Natl. Acad. Sci. USA 99, 10599-10604.

Orlic D, Kajstura J, Chimenti S, et al. Bone marrow cells regenerate infarcted myocardium. Nature 2001, 410:701-705.

Oz MC, Argenziano M, Catanese KA, et al. Bridge experience with long-term implantable left ventricular assist devices. Are they an alternative to transplantation? Circulation. 1997; 95:1844-1852.

Panis, B. and Lambardi, M. 2005. Status of cryopreservation technologies in plants (crops and forest trees).

Papa, R. 2005. Gene flow and introgression between domesticated crops and their wild relatives.

Park, T.S., Hong, Y.H., Kwon, S.C., Lim, J.M., and Han, J.Y. 2003. Birth of germline chimeras by transfer of chicken embryonic germ (EG) cells into recipient embryos.

Park, T.S., Hong, Y.H., Kwon, S.C., Lim, J.M., and Han, J.Y. 2003. Birth of germline chimeras by transfer of chicken embryonic germ (EG) cells into recipient embryos.

Peterson, R.K.D., Arntzen, C.J. 2004. On risk and plant-based biopharmaceuticals. Trends in Biotechnology 22:64-66.

Pfeffer JM, Pfeffer MA, Fletcher PJ, Braunwald E. Progressive ventricular remodeling in rat with myocardial infarction. Am J Physiol 1991; 260:H1406-14.

Pfeffer MA, Braunwald E, Moye LA, et al. Effect of captopril on mortality and morbidity in patients with left ventricular dysfunction after myocardial infarction. Results of the survival and ventricular enlargement trial. The SAVE investigators. N Engl J Med 1992;327:669-677.

Pitt B, Segal R, Martinez FA, et al. Randomised trial of losartan versus captopril in patients over 65 with heart failure (Evaluation of Losartan in the Elderly Study, ELITE). Lancet 1997;349:747-752.

Pittenger MF, Mackay AM, Beck SC, et al. Multilineage potential of adult human mesenchymal stem cells. Science 1999, 284:143-147.

Pouzet B, Ghostine S, Vilquin J-T, Garcin I, Scorsin M, Hagege AA, Duboc D, Schwartz K, Menasche P (2001) Is skeletal myoblast transplantation clinically relevant in the era of angiotensin-converting enzyme inhibitors? Circulation 104:I223-228.

Primmer, C. 2005. Genetic characterisation of populations and its use in conservation decision making in fish.

Purdue Agricultural Biotechnology. 2004. Moving Genes into Plants. http://www.agriculture.purdue.edu/agbiotech/genetransfer.html

Pursley, W. H., Presentation at the September 4, 2003, meeting of the President's Council on Bioethics, Washington, D.C., available at www.bioethics.gov.

Rafii S, Shapiro F, Rimarachin J, Nachman R, Ferris B, Weksler B, Moore AS, Asch AS. Isolation and characterization of human bone marrow microvascular endothelial cells: hematopoietic progenitor cell adhesion.

Rao, N.K. 2004. Plant genetic resources: Advancing conservation and use through biotechnology. African Journal of Biotechnology, 3, 136-145.

Ravichandran LV and Puvanakrishnan R. In vivo labeling studies on the biosynthesis and degradation of collagen in experimental myocardial myocardial infarction. Biochem Intl 1991;24:405-414.

Reich,D. E. & Lander, E. S. On the allelic spectrum of human disease. *Trends Genet.* **17,** 502–510 (2001).

Rohwed*el, J., et al., "Embr*yonic stem cells as an in vitro model for mutagenicity, cytotoxicity, and embryotoxicity studies: present state and future prospects," Toxicology In Vitro 15: 741-753 (2001).

Ryan, E. A., et *al., "Clinica*l outcomes and insulin secretion after islet transplantation with the Edmonton protocol," Diabetes 50: 710-719 (2001).

Xu, R. H., et a*l., "BMP4 init*iates human embryonic cell differentiation to trophoblast," Nature Biotechnology 20: 1261-1264 (2002).

Zalzman, M., *et al., "Rev*ersal of hyperglycemia in mice using human expandable insulin-producing cells differentiated from fetal liver cells," Proceedings of the National Academy of Sciences of th*e United States of America 100: 7253-7258 (2003).*

Zhao, M., et al., *"Ameliora*tion of streptozotocin-induced diabetes in mice using human islet cells derived from long-term culture in vitro," Transplantation 73: 1454- 1460 (2002).

Zulewski, H., *et al., "Mul*tipotential Nestin-Positive Stem Cells Isolated From Adult Pancreatic Islets Differentiate Ex Vivo Into Pancreatic Endocrine, Exocrine and Hepatic Phenotypes," Diabetes 50: 521-533 (2001).

2

Institution Related to Biotechnology Information

National Centre for Biotechnology Information

Established in 1988 as a national resource for molecular biology information, NCBI creates public databases, conducts research in computational biology, develops software tools for analyzing genome data, and disseminates biomedical information—all for the better understanding of molecular processes affecting human health and disease.

NCBI at a Glance

Mission

See how NCBI contributes toward the NIH mission of "uncovering new knowledge".

Programs and Activities

A little about NCBI's basic research program and the tools and technologies we develop and disseminate.

Organizational Structure

The branches that make up the NCBI, along with a description of their function and the function of our Board of Scientific Counselors.

Research

The basic research projects conducted by our intramural investigators.

A Story of Discovery

From sequence to survival: the search for understanding disease.

Topic

Bioinformatics

Learn why biology is being transformed from a purely lab-based science to an information science as well.

Genome Mapping

Genome mapping. What does it really mean?

Molecular Modeling

See how researchers use computers for predicting protein structure from gene sequences.

SNPs

What are SNPs and why are they important?

ESTs

Learn how ESTs can provide researchers with a quick and inexpensive route for gene discovery.

Microarray Technology

This developing technology is revolutionizing the way scientists examine gene expression.

Molecular Genetics

What is a Cell: The basics on cell organization and cell division.

What is a Genome: Every living organism has a genome. Find out why!

Molecular Genetics: Learn about the different methods that scientists use to obtain and study DNA.

Pharmacogenomics

Pharmacogenomics is making it possible to predict how a person will react to a drug based upon their genetic profile.

Phylogenetics

Learn how scientists reconstruct evolutionary pathways using computerized applications.

Databases and Tools

NCBI Resource Index

An index of all of NCBI's publicly available resources with links to each.

Human Genome Resources

Guide to Online Resources

Quick access to the resources used by NCBI scientists to piece together and analyze the human genome.

NCBI Genome Analysis Pipeline

A description of NCBI's contig assembly and annotation process.

Model Organisms Guide

Guide to Model Organisms

An Online guide to various model organisms.

Outreach and Education

Courses and Tutorials Offered by NCBI

Getting Started

A quick "how-to guide" on NCBI data mining tools, designed for the novice user.

A Field Guide to NCBI Resources

A Field Guide to NCBI ResourcesA hands-on workshop covering NCBI's molecular biology resources.

MLA Course

An introduction to NCBI's molecular biology search systems, including Entrez and BLAST.

BLAST Information Guide

An online tutorial designed to assist researchers in setting up and performing a BLAST search.

Entrez Help

An introduction to NCBI's integrated search and retrieval system.

Structure Tutorial

A guide to NCBI's visualization tool for biomolecular structures, sequences, and sequence adjustments.

PubMed Online Tutorial

A Web-based learning program designed to teach users how to search our journal literature system.

Genes and Disease

The quest for an understanding of how genetic factors contribute to human disease.

Coffee Break

A collection of short reports on recent biological discoveries.

Glossaries

BLAST

Confused by a BLAST term? Consult the BLAST glossary.

Genetics

The National Human Genome Research Institute's glossary of genetic terms.

Genome

A glossary of genomic terms from the Department of Energy's Human Genome Program.

Recommended Links

NIH Office of Science Education

Focal point for science education activities at the NIH.

Bioinformatics at the NIH

A source for information about biomedical computing at the NIH.

NIH Grants and Funding

Leads to NIH funding opportunities.

NIH Education and Training

Quick access to various NHGRI education and training activities.

The Human Genome Project: Exploring Our Molecular Selves

A multi-media educational kit designed to help the public understand how genomics will improve health.

Exhibit Schedule

NCBI staff will be available for consultation at various scientific meetings throughout the year.

From Green Revolution to Gene Revolution

The world's population is expected to exceed 8 billion by 2025 — an increase of 2.5 billion over the current level. Much, but not all, of the increase will occur in developing countries, and there will be many more mouths to feed in complex circumstances. Nobel laureate Norman Borlaug, father of the Green Revolution, calculates that "to meet projected food demands, by 2025 the average yield of all cereals must be 80 percent higher than the average yield in 1990." These increases must come primarily from increasing biological yields, not from area expansion and more irrigation; overconsumption and waste in rich countries and population pressure in poor countries have already placed dangerous burdens on the ecosystems on which we all depend. Meanwhile, poverty and hunger persist in our world of plenty despite the enormous burst of output and productivity, the dazzling changes shaped by science and technology, and the amazing achievements recorded on the social indicators for so many of the people on the planet. Food production capacity is widespread and substantial, yet millions are too poor to provide for their essential needs. Some 40,000 people die from hungerrelated causes every day.

Meeting the Global Agricultural Challenge

Agriculture is central to our management of these problems in the new millennium. Agricultural growth is essential to economic growth in most developing countries. Very few low-income countries have achieved rapid nonagricultural growth without corresponding rapid agricultural growth. Conversely, most of the developing countries that grew rapidly during the 1980s experienced rapid agricultural growth in the preceding years. Agriculture, moreover, is the primary interface between people and the environment. So agricultural transformation will be essential to meet the global challenges of reducing poverty, feeding the world's burgeoning population, and protecting the environment. This transformation will have to occur at the level of small-holder farmers so that their complex farming systems can be made more productive and efficient in the use of resources.

The challenge is both technological (requiring the development of new high-productivity, environmentally sustainable production systems) and political (requiring policies that do not discriminate against rural areas in general and against agriculture in particular), and it will have to be accomplished at a time when attention to agricultural development and rural well-being is diminishing. An essential aspect of the response to this challenge is to harness all instruments of sustainable agricultural growth.

The Role of CGIAR

A major responsibility falls on the Consultative Group on International Agricultural Research (CGIAR), the only organization in the world that exists solely to mobilize the best in agricultural science on behalf of the world's poor and hungry. The CGIAR, created in 1971, is an informal association of 58 public and private sector members supporting 16 international agricultural research centers. The United States was a founding member of the CGIAR and plays an important role in formulating its policies.

The international centers develop advanced breeding material for adoption and use by national agricultural research systems (NARS) in developing countries. The CGIAR works with a range of partners in the public and private sectors. Its research products are international public goods unconditionally available to poor farmers, national programs, and other users.

The CGIAR is ideally positioned to address the next compelling challenge that agricultural scientists must confront: combining conventional research with the promise of the genetic revolution. Just as the Green Revolution has fed millions and served as the basis of economic transformation, we have to ensure that the gene revolution leads to a "doubly green revolution" in which increased productivity and natural resource management are in balance. The poor will thereby be enabled to begin their ascent from poverty.

The Promise of Biotechnology

The revolution in the biological sciences — molecular genetics, informatics, and genomics research — has opened up all sorts of possibilities. The promise of biotechnology as an instrument of development lies in its capacity to improve the quantity and quality of plants and animals quickly and effectively. The time required to identify and combine favorable traits through traditional crop breeding is greatly reduced. Increased precision in plant breeding translates into improved predictability of the resulting products in performance and survival.

The application of biotechnology can create plants that are more drought resistant, more salt tolerant, more resistant to pests — without pesticides. Plant characteristics can be genetically altered for earlier maturity, increased transportability, reduced post-harvest losses, and improved nutritional quality. Vaccines against diseases afflicting livestock are already important products of biotechnological research.

In the past few years, there have been continuing increases in the area planted to transgenic crops. In 1998 the global area of transgenic crops more than doubled over that of 1997, with the United States leading the field with 20.5 million hectares (one hectare is equal to 2.47 acres), 74 percent of the global area. The five principal crops grown are soybeans, maize (corn), cotton, canola/rapeseed, and potatoes. In terms of transgenic trait, the largest area was occupied by herbicide-tolerant crop varieties (71 percent), followed by insect-resistant varieties (28 percent).

Most of the early products of agricultural biotechnology focus on crop protection. In 1998, transgenic crops that are herbicide tolerant covered about 19.8 million hectares. Use of herbicide-tolerant varieties greatly facilitates weed control using certain types of herbicide. It also enables farmers to employ soil conservation practices such as minimum tillage, which reduces soil erosion. As for increased plant resistance to pests, in 1998 an estimated 7.7 million hectares were planted to transgenic crops with introduced genes that produce substances toxic to target insect pests. This has resulted in the reduced use of insecticides, a positive impact not only on farm income but also on the environment. Research is under way, as well, to improve the nutritional quality or value of some food crops in developing countries. For instance, Swiss scientists have developed genetically modified rice that has higher levels of Vitamin A and iron. With some 2.4 billion people consuming rice as their staple cereal, this "new" rice can potentially prevent cases of blindness and anemia, particularly among millions of children in developing countries.

Issues and Concerns

The revolution in the biological sciences has both promise and problems. We are confronted by profound ethical and safety issues, complicated by the issues of proprietary science. Some of the concerns come from scientists who wonder if the results of these scientific efforts could produce "super weeds" or "super viruses." Many protests have been made by civil society institutions on ethical or ecological grounds. The dominance of the private sector in the North, where the bulk of developments in agro-biotechnology have so far taken place, raises fears that this will create a new phase of comparative disadvantage and increased dependency in the South.

Very much at issue are patenting and intellectual property rights (IPR). Supporters of patenting point out that if the private sector is to mobilize and invest large sums of money in research and development in agro biotechnology, it has a powerful claim to protecting and recouping what it has put into the exercise. On the other side of the argument is the fear that patenting and the exercise of IPR will lead to a monopolization of knowledge, restricted access to germplasm, controls over the research process, selectivity in the focus of research, the development of science and technology apartheid, and, thereby, the increased marginalization of the majority of the world's population.

These concerns cannot and must not be ignored. In October, the CGIAR and the U.S. National Academy of Sciences held an international conference ("Ensuring Food Security, Protecting the Environment, and Reducing Poverty: Can Biotechnology Help?") to examine the full range of issues connected with the development and use of agro-biotechnology, and in particular to discuss safeguards against its perceived hazards. The conference, held at the World Bank in Washington, brought together scientists, governments, the civil society, and professional communicators for an open discussion of the issues. The critical issue is that every instrument of agricultural transformation should be mobilized in our efforts to feed the hungry, help the poor, and protect the environment. We cannot accept the notion that deprivation is imprinted on the genes of the poor and destitute, with misery their inevitable destiny. The ethical dimension of depriving them of the advantages that biotechnology with adequate safeguards can bring must be weighed against other ethical concerns. Both sets of issues need to be confronted boldly. We must find ways of realizing the promise of biotechnology while avoiding the pitfalls.

U.S. Farmers Adopting Genetically Modified Crops

U.S. farmers have rapidly adopted genetically modified

crops since their introduction over the past few years. The benefits to growers have been higher yields, lower costs, and ease of management. An additional impact of these varieties is the potential to reduce pesticide use. While acreage planted to these new varieties has increased, so too has public awareness of and opposition to the growing prevalence of these crops in the food supply, both in the United States and abroad. In response, demands for mandatory labeling of genetically modified crops destined for Europe and Japan has been met with recent requests by U.S. food processors such as Archer Daniels Midland for growers to segregate their genetically modified crops from conventional varieties.

The additional costs of product segregation for genetically modified crops may cut into potential profits, and growers may choose not to plant genetically modified varieties in the coming years. Three major crops in the United States have been planted to genetically modified varieties: corn, cotton, and soybeans. A newly available potato variety has also been planted on a limited acreage. The adoption rates for these crops are given in the accompanying table. Each of these crops delivers a unique set of benefits to growers who adopt them, depending on pest control issues particular to the crops and whether other effective and affordable pest control options are available. This article illustrates the benefits of the genetically modified crops by examining three of the cases in which the technology has been introduced: Roundup Ready soybeans, Bt corn, and Bt cotton.

Roundup Ready Soyabeans

Roundup Ready soybean varieties were introduced in 1996. The main advantage of Roundup Ready soybeans is the simplicity of weed control, which relies on one post-emergence herbicide to control a broad spectrum of weeds without crop injury or crop rotation restrictions. Before the introduction of Roundup Ready soybeans, growers generally applied a pre-emergence herbicide before or at planting to control early season weeds, followed by either mechanical cultivation or selective post-emergence herbicides applied over the growing crop. The use of post-emergence herbicides has been increasing since these materials became available in the 1980s. This has facilitated the adoption of conservation tillage practices and narrow row spacing, which limit the opportunity for cultivation.

Although soybean growers have many post-emergence herbicide options, none has the broad spectrum of weed control of Roundup. Further, many conventional herbicides cause injury to the crop, including stunted growth, while Roundup may be applied over Roundup Ready varieties at any stage of growth without causing damage. Crop injury may not reduce yield, but it can delay canopy closure and increase weed competition with the crop.

Another limitation to conventional programs is the development of herbicide-resistant weeds. Several commonly used conventional soybean herbicides share a common mode of action, and weed populations resistant to these materials have developed in many areas of the U.S. Midwest. Finally, some of the herbicides that are used in soybeans can cause damage to subsequent rotation crops due to the persistence of the materials in soil. There are no crop rotation restrictions using Roundup, as it has no residual activity.

The costs of Roundup Ready soybeans compared to conventional varieties are expected to be roughly the same, although little field-level data exist for comparison.

Growers may have realized cost savings by switching from a program using more than one herbicide treatment to a single application of Roundup. In 1998, a program using one application of Roundup over Roundup Ready soybeans cost under $20 an acre, compared to a conventional program using a combination of herbicides that cost around $30. However, the prices of competitive herbicides have been reduced to make alternative programs more competitive with the Roundup Ready system.

Yields of Roundup Ready soybeans are expected to be about the same as conventional varieties. Although state university variety trials have shown that the yield potential of the Roundup Ready varieties has been lower than the highest yielding conventional varieties, this lag is narrowing and is expected to disappear as the herbicideresistant trait is introduced into the best conventional varieties.

Bt Corn

Field corn varieties altered to express an insecticidal protein from the soil bacterium *Bacillus thuringiensis* were also introduced for the 1996 growing season. Bt sweet corn varieties were introduced for the 1999 season. The main advantage of the Bt field corn varieties has been increased yields. For sweet corn, which is treated much more frequently with insecticides, the impact is expected to be a dramatic decrease in insecticide use. The Bt protein is selectively active against lepidopteran, or caterpillar, insects. Bt corn varieties have provided control of one of the major insect pests in corn, the European Corn Borer. Due to the difficulty in scouting for this pest and the importance of timing insecticide application before the caterpillar bores into the corn stalk and is protected from insecticides, it is estimated that less than 5 percent of field corn acreage in the U.S. “Corn Belt” was being treated with insecticides for the European Corn Borer prior to the introduction of Bt corn varieties.

Researchers have sought alternative methods to control the corn borer, although none has proven effective on a wide scale. Traditional breeding efforts to select for natural resistance to the corn borer resulted in the development of varieties with intermediate levels of resistance that were widely used into the mid 1970s. However, acreage planted to these varieties decreased dramatically due to the introduction of much higheryielding susceptible hybrids, which could produce higher yields than the resistant hybrids, even after sustaining high levels of damage from the corn borer.

Another extensive research program was undertaken by the U.S. Department of Agriculture (USDA) to identify natural predators of the corn borer. The program resulted in the introduction of 24 species of parasites in U.S. corn production areas, of which only six became established. While these beneficial insects have provided control of the corn borer in some years in some areas, their impact has been limited. Losses due to the corn borer vary each year with infestation levels, and they are generally unpredictable from one year to the next. Largely uncontrolled until the introduction of Bt corn varieties, the European Corn Borer has caused production losses that have ranged from 33 million bushels to over 300 million bushels per year.

Bt corn varieties have been shown to provide a very high level of protection from the corn borer. Plants sustain only minute damage as the corn borer larvae attempt to feed. The benefits that growers will realize from planting Bt corn varieties depend on the level of infestation. In light infestation years the benefits may not be great, while in heavy infestation years growers will realize substantial yield increases. An average of available research results comparing yields from Bt and non-Bt corn fields indicates that growers experienced a yield advantage of approximately

12 bushels an acre (2.47 acres are equal to one hectare) in 1997 and four bushels an acre in 1998.

The price premium for Bt corn was approximately $10 an acre in 1997 and 1998. Assuming corn prices of $2.43 a bushel in 1997 and $1.95 a bushel in 1998, the average income changes for Bt corn were an increase of $18 an acre in 1997 but a decrease of $1.81 an acre in 1998. It is expected that in 10 of the 13 years from 1986 to 1998, corn borer infestations in the Corn Belt were such that corn growers would have realized a gain from planting Bt corn.

Bt Cotton

Bt cotton varieties were introduced in 1996 and were adopted very rapidly in some cotton production areas. Alabama growers planted Bt cotton varieties on 77 percent of the state's cotton acreage in the first year the new technology was available. The U.S. adoption figures (17 percent in 1998) mask what are actually high adoption rates because some major production areas, such as Texas and the San Joaquin Valley of California, do not face infestations of the pests the Bt cotton varieties control and do not need the technology. The primary benefit to growers of adopting Bt cotton varieties is controlling insect pests that have grown resistant to commonly used insecticides. Growers have also reduced the number of insecticide treatments and insecticide treatment costs. In 1998, 71 percent of cotton acreage was treated with insecticides to control pests that would otherwise reduce yields. In most states, over 90 percent of the acreage was treated, while in Texas only 47 percent was treated due to low pest pressure in some areas. Bt cotton varieties control three of the primary insect pests of cotton: tobacco budworm, cotton bollworm, and pink bollworm. The Bt cotton varieties are most effective in controlling tobacco budworm and pink bollworm and provide lesser control of cotton bollworm due to its greater ability to survive on more mature cotton plants. Budworm and cotton bollworm cause yield losses in most cotton production areas of the United States outside of California and some areas of Texas. The pink bollworm is prevalent in southern California production areas, as well as Arizona, New Mexico, and far west Texas. Several of the commonly used insecticides for these cotton pests are pyrethroids, which have provided low-cost, effective control. However, in many areas, pyrethroid-resistant populations of tobacco budworm have developed, diminishing the usefulness of these insecticides. In 1995, a year with high tobacco budworm infestation in many areas, Alabama growers sustained nearly 30 percent yield losses due to the ineffectiveness of the pyrethroids in face of high levels of budworm resistance. Growers who adopt Bt cotton varieties have been able to achieve effective insect control of the three target pests with a reduced number of insecticide applications. USDA pesticide use data show a reduction of 2 million pounds of the insecticides that are recommended for the control of these insects since the introduction of Bt cotton varieties. Insect control costs, including the $32- per-acre technology fee, are expected to be lower than for conventional varieties. In addition, growers in boll weevil eradication zones have adopted Bt varieties at a higher rate than in other areas in order to achieve control of bollworm/budworm in the face of treatments that destroy their natural predators. It is expected that growers will also experience increased yields using Bt varieties compared to conventional varieties.

The reasons U.S. growers are adopting genetically modified crops vary. The fact that new genetically modified varieties are capturing a large portion of the very competitive pesticide markets for major crops indicates that the new technology is delivering benefits to growers. Understanding these benefits is critical in an evaluation of the impact that the introduction of this technology has had on U.S. agriculture.

Soybean growers have adopted Roundup Ready soybeans largely due to the simplicity and efficacy of weed control provided by the program. Bt corn varieties have provided insect control for a destructive insect pest that was mostly uncontrolled previously, which has resulted in increased yields. Cotton growers have been able to control insect pests with fewer insecticide treatments. However, the future of marketing for genetically modified crops remains uncertain. Much depends on the extent of demand for separate marketing channels for nongenetically modified crops and who bears the costs of identity preservation. If marketing agents demand that farmers limit the mixing of these crops with those that have been genetically modified, costs will rise and the benefits farmers have realized using this new technology may disappear.

BIO-Patent Revolution

The persistence of hunger and the accompanying extreme poverty found in many developing economies are indefensible during a period of rapid advances in food production using agricultural biotechnology. The genetically engineered wheat gene, Norin 10, for example, has helped such countries as India and Pakistan increase their wheat harvests by 60 percent. Similarly, Costa Rican scientists have created a new genetically modified rice, one that will help not only Costa Ricans but should be suitable for India, Vietnam, Japan, and other rice-growing countries as well. Biotechnology offers the promise of crops that not only generate their own pesticides to create disease-free products, but also plants with genes that provide improved nutritional content and allow the products to stay on the market shelf longer without rotting.

At the same time, developments in genetically engineered foods are occurring in a proprietary environment, where biotechnology products and the processes are being patented and their release to the public restricted because of the enormous investments by the private sector. Should Costa Rica have exclusive rights to its newly created rice, or should it be shared freely with other countries? Who owns the rights to nature and that which nature produces? There are no easy answers. The challenge facing society is how to advance agricultural science without jeopardizing intellectual property rights and thereby guarantee greater food security for emerging economies. This article seeks to raise some of the key questions that policy-makers and courts must address and whose outcome will have a profound influence on the development and trade of biotechnology products.

Encouraging Investment in Biotechnology

Incentive is vitally important for any innovation that requires a large expenditure of funds and intellectual innovation. To take a bioengineered product out of the laboratory and onto the fields as a commercial crop, the greatest possible care must be exercised to test for the safety of the environment. Large numbers of clinical and field trials are hugely expensive and lengthy, but necessary. The international patent regime is thrust squarely in the middle of the debate because, without an effective international patent regime, there is no realistic possibility that the private sector will invest the significant resources required to advance state-of-the-art genetic engineering.

Without patent protection that would exclude others for a limited period of time, once a product hit the market it could easily be duplicated by everyone. This is particularly true when dealing with "living" inventions. If there is a new strain of rice, all that would be needed to copy the invention and enter into competition with the originator would be a small investment in a modest amount of seed and then the collection of the seed from the resulting crop. Clearly, if the originator of a genetically engineered crop can have no competitive advantage, then what incentive is there to spend the millions of dollars necessary to develop and market a new genetically engineered product? The reality is that without a patent regime, there would be far fewer new products that have the potential to better the human condition. At the same time, patent law must be structured in a way that not only encourages the original inventor but also provides access to others so that the new product can be improved upon. And it must protect against commercial concerns that obtain patents in the area of biotechnology without carrying through with a concrete product, thus depriving society of a direct benefit.

Who Owns an Invention?

Should, however, a patent be granted to something found in nature? One could very well imagine the outcry that would be created if someone enters a rain forest, pulls out a leaf with special medicinal qualities, and then seeks a patent on that leaf. What "invention" was made on the part of man? If the leaf is already known as a native medicinal, then there may not be anything patentable. To be patentable, something must be "new." If someone discovers that a purified extract from that leaf has medicinal qualities, the extract may be patentable.

Indeed, Nobel Laureate Sune Bergstrom's isolation of prostaglandin, a class of molecules that help in the control of blood pressure, did not make him the "inventor" of prostaglandin; rather, he was the inventor of a "purified" prostaglandin that had not previously existed in nature. While someone clearly may be determined to be the "inventor" of a form of the leaf, who owns that invention? A country is free to legislate ownership rights as it sees fit to meet its national policy goals. If the inventor works for a company, in most cases the patent right will go to the company rather than to the inventor.

But merely to discover the active principle involved in the medicinal qualities of the purified form of our leaf does not end the matter of patent ownership. Take, for example, the case of steroids. Decades ago, scientists discovered a basic steroid molecule with its characteristic arrangement of carbon rings. Scientists created synthetic steroids by manipulating chemicals to introduce various atoms onto the steroid's molecular structure. Each of these synthetic modifications, whether made in California or Kyoto, is considered a separate invention and not part of the original invention of the synthetic steroid.

Just as the discovery of the basic steroid molecule has limited benefit to the patent owner, the discovery of a purified leaf extract has limited benefit to the inventor. The real value of the discovery is bestowed on those who produce modifications that lead to needed and valuable products. Since the knowledge of the structure of that purified leaf extract will be in the public domain as part of the scientific literature (and since anyone is also free to use that structure for research even if it is patented), patent law provides little benefit to the "leaf" patent owner unless the structure of the leaf is kept secret. Then the owner of the invention, whether an individual or a company, will have the lead time to first identify the key structure of the purified leaf extract and be able to obtain patents on that structure.

Technology Transfer

Obviously, inventions cannot be developed everywhere.

Does a company that has the patent on the leaf structure give away the rights to other countries? As I previously stated, that would reduce, and in most cases probably eliminate, the incentive for developing the new product. Consider, for example, a Latin American development of a new crop line that would be suitable in Latin America, Japan, and elsewhere. If the Latin American inventor has not exploited his invention in foreign markets, other countries may be able to take advantage of the technology by providing certain compensations to the investor. Such compensations might include royalty income or an agreement to expend additional resources to improve the initial technology — the possibilities are limitless.

Not all countries have the resources to buy into the new technology. Here there is a role for international development and research organizations such as the Consultative Group on International Agricultural Research and the World Bank. These groups need to consider whether they should allocate additional resources toward the development of genetically engineered products that would be freely shared with the poorest countries.

The Importance of Patent Protection

A direct result of the Uruguay Round global trade accord concluded several years ago was the establishment of minimum standards for patent and related intellectual property protection, including those for products resulting from genetic engineering. Such standards are vital not only for protecting products from industrial countries but also for encouraging the creation of and protecting developing country inventions. Countries such as India, with its large number of highly trained scientists and engineers, are well placed for taking advantage of patent protections. Brazil and China are examples of countries that have made great strides in both the creation and implementation of modern patent laws. Those countries that follow the lead of China and Brazil will have the incentives of the patent system to fuel innovation. Those that fail to move forward with strong patent laws will remain on the sideline, trapped by the lack of an effective patent law, unable to provide the legal setting to foster and encourage innovations in the local economy.

Safety and Choice

Genetically engineered (GE) foods and food ingredients are widespread in our diet — from infant formula to corn-muffin mix to McDonald's McVeggie Burgers, according to testing in the United States by the publication *Consumer Reports*. Meanwhile, news reports worldwide tell of consumers, farmers, and activists dumping milk, suing governments, destroying GE crops, and persuading supermarket chains and companies to shun GE ingredients.

Why all the fuss? In part, it's fueled by attitudes among regulators and manufacturers that dismiss unanswered questions about the technology; ignore ethical and societal issues, pretending that only science matters; insist that GE foods are "equivalent" to conventional foods, when consumers can and do make distinctions between them; and bully consumers or farmers (for example, through lawsuits designed to stop dairies from disclosing that their cows are not treated with BGH/BST, a genetically engineered hormone).

Sharing power with those affected by biotechnology and addressing their legitimate concerns would help diffuse the controversy; making the risks more voluntary would make them seem smaller.

Consumer Concerns

Here are some of the issues that matter to consumers.

Allergenicity: Allergic reactions to foods are hard to

predict, but they can be life-threatening. Once sensitized, individuals may react more strongly to subsequent exposures to the same allergen. According to the report of a joint Food and Agriculture Organization/World Health Organization Expert Consultation on Biotechnology and Food Safety in 1996, more than 160 foods are associated with sporadic allergic reactions, with children at greater risk than adults.

Genetic engineering can introduce unknown allergens into food. Virtually every gene transfer in crops results in some protein production, and proteins are what trigger allergic reactions. Biotechnology can introduce new proteins into food crops not just from known sources of such common allergens as peanuts and shellfish, but from plants of all kinds, animals, bacteria, and viruses whose allergenicity is largely unknown. If someone has an allergic reaction to a conventional food, he or she can check the food label, which typically identifies all the ingredients precisely enough that the person can avoid future re-exposure to the allergen. However, if one has a reaction to an allergen in a food on account of a genetic modification, and the label does not disclose the presence of the GE food, one has little clue as to what to avoid in the future. This could very well increase the risk of subsequent exposures to the allergen. Or it could to lead to people needlessly avoiding foods that are actually safe for them.

Environmental Risks: In a study conducted by researchers at Cornell University in the United States, pollen from corn genetically engineered to contain Bt toxin (a toxin from a soil bacterium that kills insect pests) can kill monarch butterfly caterpillars, at least in the laboratory. This recent finding adds to other studies suggesting adverse effects on beneficial insects from genetically engineered plants. This has boosted environmental concerns over genetic engineering, inasmuch as insects, among their multiple roles in living systems, are key pollinators of many plants. Genetically engineered genes might also accidentally cross over to non-target plants. For example, "terminator" genes might cross over and cause plants to produce sterile seeds, resulting in the significance loss of food and diversity. Similarly, genes from plants engineered to be herbicide resistant could cross over to other plants, creating "superweeds." Studies in Norway and the United States have already demonstrated that the gene for herbicide resistance can move from cultivated canola to wild relatives.

Ethical/Religious Considerations: Some people find genetically engineered foods unacceptable for ethical or religious reasons. Rabbis, ministers, vegetarians, and others can be found on both sides of the debate regarding foods engineered to contain genes from animals or species that are proscribed by certain religions. Labeling allows those consumers to choose according to their conscience without imposing that view on others.

Social/Economic Justice Issues: Many consumers are suspicious of who is controlling a technology that promises to revolutionize agriculture. Biotechnology enables agricultural production to become more vertically integrated, consolidated, and centralized, largely in the hands of multinational corporations. As author and professor Joan Gussow says, "Someone is going to produce and subsequently manipulate the materials out of which each of us is made. Are we really prepared to trust that responsibility to Phillip Morris?" Probably no GE product has generated as much controversy as the so-called terminator technology, which makes plants produce sterile seeds. Particularly in a country such as India, where farmers routinely save and replant seeds, the prospect of being forced to buy seeds each year from large-scale corporations has generated considerable outrage, leading groups of farmers to burn plots where the seeds were being studied.

Unanticipated Consequences: No technology, no matter how beneficial, is risk free. And with any new technology, there may be unanticipated consequences. The recent revelation that pollen from plants genetically engineered to contain Bt may harm monarch butterflies is a case in point. Genetic engineering could also bite back by exacerbating the growing problem of antibiotic resistance. Most genetically engineered plants contain a gene for antibiotic resistance as an easily identifiable marker. Hypothetically, antibiotic resistance genes may move from a crop into bacteria in the environment and — since bacteria readily exchange antibiotic resistance genes — move into diseasecausing bacteria. In September 1998, the British Royal Society called for ending the use of antibiotic resistance marker genes in engineered food products. Many consumers know nothing about antibiotic resistance genes used in GE foods, but they do understand the unpredictability of complex systems. Genetically engineering a plant to have a particular trait can have unexpected ripple effects on the ecosystem that cradles it. It is consumers' grasp of this fundamental phenomenon that underlies much of the concern over biotechnology.

Misperceptions of Consumer Reaction

GE foods per se do not worry many consumers as much as the covert manner by which they have crept into the marketplace. Here are some common misperceptions I frequently encounter.

Misperception No. 1: Most consumer organizations are anti-technology and oppose all genetically engineered foods. In fact, many independent consumer organizations recognize the potential benefits of biotechnology. Consumers Union, publisher of *Consumer Reports* and the largest and oldest consumer organization in the United States, recognizes that, with proper safeguards, genetic engineering offers the possibility of foods that could benefit consumers. Consumers International, a federation of over 230 consumer organizations from some 100 countries, is pro-labeling, not anti-biotechnology. The problem is that while consumers have been promised foods that taste better, are healthier, and will help "feed the world," the applications of the technology to date have fallen flat in delivering consumer benefits. BGH/BST, a genetically engineered drug that makes cows produce more milk, has not translated into lower prices. The main benefit of Roundup Ready soybeans and cotton has been to expand the market for Roundup. Bt corn, cotton, and potatoes are spreading pest resistance to this natural and safe insecticide, as well as threatening beneficial wild insects. And most consumers know that there is no such thing as a "magic bullet" solution to the complex problem of world hunger, which has less to do with agricultural production capacity than with political priorities and the distribution of economic resources.

Misperception No. 2: Labels on genetically engineered foods will be viewed by consumers as warning labels. The argument that labeling will stigmatize GE foods in the consumers' eyes as potentially unsafe assumes that only science matters, and that the public is too ignorant to understand the data. Consumers need labeling to make an informed choice based on whatever criteria they wish to use. Labeling with full disclosure would be a positive step toward a more informed citizenry and a way to increase consumers' familiarity with this new technology.

Misperception No. 3: U.S. consumers, unlike consumers in Europe and many other countries, are comfortable with GE foods and don't think labeling is necessary. Actually, only about one-third of Americans interviewed were aware that GE foods are available in the supermarket, according to a recent survey by the International Food Information Council. But many of those who are informed are not going along quietly. Earlier this year, a coalition sued the U.S. Food and Drug Administration (FDA) to block the use of a dozen GE crops as an "imminent" threat to the environment. Last year, a coalition of scientists, religious leaders, health professionals, and chefs filed a lawsuit against the FDA, claiming that the failure to label violates the agency's mandate to protect the public's health and provide consumers with relevant information about the food they eat.

Right to Know, Right to be Informed, Right to Choose

The fundamental purpose of ingredient labeling is to tell consumers what is in the foods they buy. For people subject to food allergies, that information is essential to their health. In the United States, regulators argue that the method by which a plant is developed is not "material" information in the sense of the law. Consumer organizations object that this narrow interpretation gives no weight to material facts that consumers consider important, and that it substitutes the value judgments of regulators for those of consumers. They don't agree with the "it's the product, not the process, that matters" argument. Surely, it would be considered "material" if experimentation on living human subjects was part of the development of crashworthy sports cars. Survey after survey around the world shows that while consumers are generally not opposed to genetically modified foods, they are against allowing such foods to be sold without adequate labeling. Let the success of biotechnology depend on whether informed consumers "vote" for it in the marketplace.

Public Perceptions

The first agricultural products of biotechnology have reached world markets. These products have received a frosty response in some parts of the world. But despite some recent sensational headlines in the press, North American markets have so far remained calm as foods containing ingredients developed through biotechnology have started arriving in stores. It is clear from a review of consumer surveys conducted in the United States, Japan, and Europe that consumer perceptions about biotech foods are strongly influenced by type of information, confidence in government, and cultural preferences.

Varying Viewpoints

In general, consumers worldwide see considerable value in human genetic testing, the development of new medicines to combat disease, and the use of biotechnology to develop new types of insect-resistant crop plants. Consumers are less likely to accept the use of biotechnology with animals (even to enhance human health), and they appear less accepting of food products developed through biotechnology, compared to crop plants, which some consumers don't even directly connect with food. The most acceptable applications are those that offer a clear consumer benefit, as well as those that are perceived to be ethical and safe.

However, public attitudes about agricultural biotechnology vary considerably among countries. Consumers from Canada, Finland, Italy, Japan, the Netherlands, Portugal, and the United States are more positive about biotechnology than most other countries. Support for biotechnology is much lower in Austria, Denmark, Germany, and Sweden. Two countries — the United Kingdom and France — used to be quite positive toward biotechnology but have become more hostile in the past year or so. The United Kingdom has become more negative for a number of reasons, including fallout from mad cow disease, antibiotechnology comments made by Prince Charles, food retailer panic, and an effective network of activist groups.

France has become more negative partly due to French farmers' opposition to American grain imports and a broader French cultural opposition to what they see as the globalization (that is, Americanization) of the food supply. Surveys in the United States (as recently as the spring of 1999) have consistently shown that between two-thirds and three-quarters of American consumers are positive about biotechnology, and about three-quarters have consistently expressed a willingness to buy insectprotected produce developed through biotechnology. One of the reasons cited for this willingness is that these products require fewer chemical pesticides. Support is highest among men and people with more formal education.

This is not to say that consumers don't have questions. Consumer groups have raised a number of concerns about agricultural biotechnology, particularly that it might somehow involve long-term or unexpected effects. Environmentalists often focus on possible ecological impacts from the use of biotechnology. While these raise important questions, they are not usually on the top of an average consumer's mind. Furthermore, consumers usually associate ethical issues with human or animal genetics, rather than with plants.

An Informed Public?

Surveys indicate fairly high awareness of biotechnology in Austria, Denmark, Luxembourg, and Sweden. But these are exceptions. In the United States, surveys since 1992 show that only about one-third of U.S. consumers have heard or read much about biotechnology — except for a brief period in 1997, when increased media attention on the cloning of a sheep raised awareness to almost 50 percent. Only about one-third of Japanese consumers reported much awareness of biotechnology in 1995 or 1998. Awareness in

France and the United Kingdom (as measured a couple years ago) was comparable to that in the United States, but it has risen for reasons discussed earlier.

Most people get their information on biotechnology from media coverage. If the media do not cover a particular story, the public tends to ignore that issue. The tone of information in the media has an important impact on consumer perceptions. Up until now, media coverage in the United States has tended to be positive and balanced (which helps account for the relatively high levels of acceptance of biotechnology).

This is a sharp contrast to the media coverage in the European Union. In fact, media coverage in the United Kingdom has taken on the characteristics of sensational tabloid journalism. The British media tend to rely on the use of emotional terms such as "Frankenfood." They also have been quick to jump on any negative allegations even when scientific consensus refutes the charge (as was the case with a controversy over the safety of insect-protected potatoes). That accounts for some of the negative consumer and food industry response in the United Kingdom.

Contributing to misinformation on biotechnology is the low knowledge in most countries of basic agricultural and biological sciences. This lack of understanding generates concern, especially when coupled with negative media coverage. There also appears to be a lack of understanding about traditional plant breeding.

Countries with the highest levels of knowledge are Canada, the Netherlands, Sweden, and the United States. Countries with the least knowledge include Austria, Greece, Ireland, Portugal, and Spain. Survey results show that providing factual information increases consumer acceptance (at least in the United States, Canada, and Japan). However, surveys also show that the source of the information may be an important factor in consumer preferences, and that a source trusted in one country is discredited in another. North American consumers have the most trust in independent health and scientific experts. In particular, acceptance increases significantly when American consumers learn that groups such as the American Medical Association, the U.S. Food and Drug Administration (FDA), and others have determined that the foods from biotechnology are safe. Japanese consumers also report high levels of trust in third-party scientific information sources. On the other hand, European consumers express the most trust in consumer and environmental groups. Their trust in government and industry is much lower than in North America.

Labeling for Whom?

The most challenging issues surrounding agricultural biotechnology involve labeling. European consumers have generally been encouraged by consumer activists to demand labels identifying foods that have been developed through biotechnology. Several food retailers (especially in the United Kingdom) have tried to exploit the public concern as a marketing tool. Europe has labeling policies in place, but they have not yet been able to establish workable regulations or procedures. They are now grappling with difficult technical issues such as which methods to use to identify traces of biotechnologyderived ingredients. They also are trying to determine what percent of ingredients in processed foods can be derived from biotechnology and still allow the food to qualify as "biotechnology-free." For the U.S. consumer, the Food and Drug Administration, an agency within the U.S. Department of Health and Human Services, has determined that a food product should be labeled as a product of biotechnology only if it has been changed in a significant way. FDA policy, supported by over 75 percent of U.S. consumers according to two national surveys, ensures product availability while providing consumers with relevant information about food safety and nutritional changes.

Recent focus groups in the United States also have demonstrated that the wording on labels has a significant effect on consumer understanding and acceptance of biotechnology. Many U.S. consumers are already overwhelmed by the level of detail on food labels and do not really want more information that has no scientific justification. Consumers want to know how a product has been changed and whether it has been approved by a government agency. Any label information needs to be simple, relevant, and clear.

The labeling of processed foods presents a number of logistical challenges and costs for everyone involved. For example, U.S. consumers have reported little need to label a bottle of ketchup that includes tomatoes developed through biotechnology in addition to traditionally bred varieties. In fact, most people don't even understand that different varieties of vegetables or fruits are currently blended during processing. In addition, consumers are not willing to pay extra to have foods labeled as a product of biotechnology (especially when this information has no meaning). Consumers want meaningful choices that are truly different. The "organic" market niche already provides a viable opportunity for consumers who do not want to consume foods developed through biotechnology, for whatever reason.

Where to From Here?

Biotechnology is at a crossroads in terms of public acceptance. Actions and statements by industry, government, and scientists over the next year will have a major influence on the long-term viability of the agricultural biotechnology enterprise. Without a major commitment to consumer education and informed choice, opposition will continue to grow. Such efforts must be based upon ongoing research into the knowledge and attitudes held by consumers and opinion leaders.

Different parts of the world clearly require different approaches. Research results suggest that biotechnology should not become a major issue for most North American consumers. Most U.S. consumers (as well as many others around the world) remain cautiously optimistic about the benefits of biotechnology. They will accept the products if they see a benefit to themselves or society and if the price is right. In fact, we are finding that consumers' responses to foods developed through biotechnology are basically the same as for any other food. Taste, nutrition, price, safety, and convenience are the major considerations. How seeds and food ingredients are produced will be relevant only for a small group of concerned "organic" consumers.

In countries where consumers are more negative about biotechnology — Germany, Austria, Sweden, and Denmark — media coverage and activist opposition have been more pronounced. In these four countries, discussion of the benefits of biotechnology have generally been ignored, while the potential risks have been emphasized. Basic social values and cultural beliefs also explain much of the differences in responses between countries. These are not necessarily amenable to educational efforts.

There also are a number of fundamental cultural differences. For example, consumers' attitudes about biotechnology are closely related to their general beliefs about science, technology, and food. European consumers tend to view farms as public natural areas where they can visit on weekends. Farms in the United States tend to be concentrated in the midsection of the country, far away from the urban population centers. Also in the United States, there has always been strong public support for and appreciation of new technology. Such support has not been as strong in parts of Europe. Some Europeans tend to view their food with an almost spiritual reverence, which is quite different from the common American view of food as fuel. These and other issues need more careful attention.

Another reason for the sustained U.S. support for biotechnology has been a long-term commitment to the education of opinion leaders and consumers. There has been an unprecedented partnership between the government, industry, universities, and third-party groups (such as the American Dietetic Association) to understand and address public concerns well before the products of agricultural biotechnology are released. There is a critical need to renew that commitment to education, information, and social science research.

Our experience in the United States provides some guidelines for a global program of information and education. Consumers need to recognize the existing benefits and future promises of biotechnology. The opportunity that biotechnology provides for feeding the world (while protecting the environment) will be compelling for many consumers. It is also important to build trust in government and scientists to serve the public interest. This requires that farmers, scientists, government officials, and others work together to ensure that consumer decisions are based on balanced information.

Monarch Butterflies and BT Corn

The publication on May 20, 1999, of preliminary research by Cornell University suggesting that the use of genetically modified corn may have toxic effects on larvae of the monarch butterfly has generated a huge amount of publicity and almost as much misinformation. Groups opposing biotechnology have used the preliminary data to argue against production and trade in all genetically engineered crops regardless of the facts. Scientists are now conducting several follow-up studies examining the effects of bioengineered corn pollen on butterflies.

While a review of current research indicates that scientists have found some risk to monarch butterfly caterpillars from Bt corn pollen, very few definitive conclusions can be made at this time, according to British biologist M.J. Crawley. Crawley observes that the data gathered to date are based on preliminary laboratory studies, carried out over a comparatively short period of time under a specific set of conditions. "They do not address the issues over the entire life cycle of the insects," he says. Numerous scientists, industry representatives, and government officials strongly support collecting more definitive data from comprehensive field research in order to resolve what has become a burning issue in the larger biotechnological debate. Following is a brief review of the Cornell preliminary research and ongoing research.

The Corell University Research

The official account of Cornell entomology professor John Losey's preliminary laboratory research was published in *Nature* magazine's May 20, 1999, issue under the title "Transgenic Pollen Harms Monarch Larvae." Losey's research investigated how pollen from genetically engineered corn (Bt corn) affects monarch caterpillars, whose exclusive food source, milkweed, frequently grows in and around corn fields. His one-page scientific correspondence in *Nature* outlined the methodology that he and two researchers used to compare the feeding, growth, and mortality of monarch larvae fed on milkweed leaves dusted either with Bt corn pollen or non-Bt corn pollen, or not dusted with pollen at all. Losey reported that the larvae "reared on milkweed leaves dusted with pollen from Bt corn ate less, grew more slowly, and suffered higher mortality than larvae reared on leaves dusted with untransformed corn pollen or on leaves without pollen." He wrote: "These results have potentially profound implications for the conservation of monarch butterflies." Noting that the amount of Bt corn planted in the United States is projected to increase markedly, Losey observed, "it is imperative that we gather the data necessary to evaluate the risks associated with this new agrotechnology and to compare these risks with those posed by pesticides and other pest-control tactics." In a Cornell University press release issued on May 19, he described his research as "just the first step" and again called for more research. Losey reports that he has been carrying out follow-up experiments and hopes to publish his latest results this winter.

ADDITIONAL RESEARCH

Another study frequently cited in the media is the ongoing fieldwork by entomologist John Obrycki and graduate student Laura Hansen at Iowa State University. While they have not yet published this research, the abstract they wrote for presentation at an Entomological Society of America meeting describes its focus and preliminary findings. Obrycki and Hansen have been investigating the potential risk that the manifestation and dispersal of Bt toxin in corn pollen poses to monarchs.

The first step of their methodology was to place potted milkweed plants at several distances from the edge in both a Bt cornfield and a non-Bt field to determine pollen concentration levels. They then took milkweed leaf samples to assess the mortality of newly hatched monarch larvae exposed to either Bt corn or non-Bt corn pollen. They found that "within 48 hours, there was 19 percent mortality in the Bt corn pollen treatment compared to zero percent on non-Bt corn pollen exposed plants." In evaluating their research, Marlin Rice, an entomologist at Iowa State, writes that both the Iowa State and Cornell studies "suggest that some, but not all, monarch caterpillars may be killed when they eat Bt corn pollen." However, he says, the bottom line is that more research needs to be conducted on the effects of Bt corn on monarchs and other non-target species. Another team of scientists from Iowa State, the Agricultural Research Service of the U.S. Department of Agriculture, and several other American universities is currently conducting field studies under the sponsorship of the Agricultural Biotechnology Stewardship Working Group (ABSWG). ABSWG is a consortium of biotechnology companies and associations that includes the Biotechnology Industry Organization (BIO), the American Crop Protection Association (ACPA), Monsanto Company, and Novartis Seeds Inc. Through ongoing projects focusing on milkweed distribution, pollen movement, monarch biology, and the biochemistry of Bt pollen, the researchers are studying how the pollen that is transported outside of corn fields affects butterfly larvae in their natural habitat, as they feed on milkweed plants. In a press release, BIO's vice president for food and agriculture states that the consortium's goal was to "assemble a cadre of top-flight, highly reputable and credible public researchers to answer the real questions that grew out the Cornell University laboratory study." According to the executive director of ACPA's Biotechnology Committee, Leah Porter, the researchers plan to present their findings this December at the Entomology Society of America's annual meeting in Atlanta. At the meeting, a symposium on the Impact of Transgenic Corn Pollen on Monarch Larvae will feature the latest results of several high-profile studies. Also participating on the ABSWG-affiliated research team are scientists from the Agricultural Research Service (ARS).

According to an ARS information officer, the USDA scientists are continuing to conduct follow-up studies and have not yet released data from these experiments. An informal meeting to exchange information and discuss the direction of future research is planned for November 2 in Chicago. In a discussion of the Cornell preliminary research, the USDA fact sheet, "USDA and Biotechnology," reports that the department is working to identify useful follow-up information and research to better understand how monarchs and corn pollen interact in the field. The fact sheet lists a number of reasons why the effect of Bt corn pollen on monarchs may prove to be small. These include corn pollen's heavy weight, which may prevent it from being blown a significant distance from the field, and the monarchs' potential capacity to avoid feeding on milkweed plants dusted with Bt pollen. Overall, the document states, "USDA is committed to further research on the potential impacts of new technologies in agriculture."

European Attitudes of Agricultural Biotech

In the face of staunch public opposition and widespread bureaucratic obstacles at the national and European Union levels, the European agrochemical industry is working hard to create a climate of acceptance for genetically engineered crops. Many agrochemical executives complain that the European Union's Commission (EC), which drafts regulations, tends to overregulate and is suspicious of new technologies.

Indeed, the European Union has fallen behind other parts of the world in research and development of genetically engineered agricultural products. Agrochemical giants such as Germany's Hoescht and Switzerland's Ciba-Geigy have been forced to move to countries such as the United States, which has a pragmatic regulatory scheme, to develop and test their products.

The current backbone of regulatory oversight in Europe for genetically engineered crop plants is the Union's "Directive on the Deliberate Release into the Environment of Genetically Modified Organisms." This regulatory process carries the organisms from the research and development level through field testing and commercialization. Before placing genetically engineered products on the market, manufacturers or importers must demonstrate that such products do not pose risks to human health or the environment.

EU directives are legally binding but require further action by member states concerning implementation and timing. For a transgenic crop plant to complete the journey from field to market in Europe, permission must be obtained first from individual countries for field trials. Depending on the country, this can take years to complete. Next an application must be made to enlarge and prepare the crop for marketing. At this stage, other EU countries can raise objections. Finally, another permit is required to actually market the product.

Recognition is dawning within the EU that if the Union is to avoid becoming simply a market for, rather than a producer of, biotech products, then regulations must be eased. EU officials now are in the process of lightening permit requirements for field tests of organisms and genetic constructs that present low risk to the environment and human health. This basically would replace permits with a notification process and reduce the time periods involved in regulatory procedures. These moves are backed by European industry, which has called for scientifically sound regulations that do not discriminate against products because of the technology used to develop them.

Not all parties in the EU are enthusiastic about the proposed changes. Germany in particular, with its strong

Green Party, may be difficult to win over. There have been four instances of vandalism against genetically engineered crops in Germany. France, the United Kingdom, and Belgium are more supportive, although the recent field trial in the UK of a virus modified with a scorpion toxin gene to enhance death rates of target insects on plants has energized opposition to the environmental release of genetically engineered organisms in the UK. In addition, a number of EU member states haven't yet fully implemented the current regulations even though the deadline for implementation was October 1991.

As one Italian official said about the proposal to ease the EU's biotechnology regulations, "We are at the beginning of the discussion on the issue. Our aim is to have the welcoming approach of the U.S."

Case Study: Stem Cells in the Future Treatment of Type- 1 Diabetes?

1. The Disease and Its Causes.

The human body converts *the sugar glucose into cell* energy for heart and brain functioning, and indeed, for all bodily and mental activities. Glucose is derived from dietary carbohydrates, is stored as glycogen in the liver, and is released again when needed into the bloodstream. A protein hormone called insulin, produced by the beta cells in the islets of the pancreas, facilitates the entrance of glucose from the bloodstream into the cells, where it is then metabolized. Insulin is critical for regulating the body's use of glucose and the glucose concentration in the circulating blood.

The body's failure to produce sufficient amounts of insulin results in diabetes, an extremely common metabolic disease affecting over 10 million Americans, often with widespread and devastating consequences. In some five to ten percent of cases, known as Type-1 diabetes (or "juvenile diabetes"), the disease is caused by "autoimmunity," a process in which the body's immune system attacks "self." T-lymphocytes attack the patient's own insulin-producing beta cells in the pancreas.

Eventually, this results in destruction of ninety percent or so of the beta cells, resulting in the diabetic state.

With a deficiency or absence of insulin, the blood glucose becomes elevated and may lead to diabetic coma, a fatal condition if untreated. Chronic diabetes, both Type-1 and the much more common Type-2 diabetes (which is not autoimmune, but largely genetic), causes late complications in the retina, kidneys, nerves, and blood vessels. It is the leading cause of blindness, kidney failure, and amputations in the U.S. and a major cause of strokes and heart attacks.

Type-1 diabetes is a devastating, lifelong condition that currently affects an estimated 550,000-1,100,000 Americans, including many children. It imposes a significant burden on the U.S. healthcare system and the economy as a whole, over and above the disabilities and impairments borne by individual sufferers. Recent estimates suggest that treatment of all forms of diabetes costs Americans a total of $132 billion per year. At 5-10 percent of all diabetes cases, the costs of Type-1 diabetes can be estimated as $6.5-$13 billion per year.

2. Current Therapy Choices and Outcomes.

The current *treatment of Type-1 diabetes consists of* insulin injections, given several times a day in response to repeatedly measured blood glucose levels. Although this treatment is lifeprolonging, the procedures are painful and burdensome, and in many cases they do not adequately control blood glucose concentrations. Whole pancreas transplants can essentially cure Type-1 diabetes, but fewer than 2,000 donor pancreases become available for transplantation in the U.S. each year, and they are primarily used to treat patients who also need a kidney transplant. Like all recipients of donated organs, pancreas transplant recipients must continuously take powerful drugs to suppress the immunological rejection of the transplanted pancreas.

In addition to treatment with whole pancreas transplantation, small numbers of Type-1 diabetes patients have been treated by transplantation of donor pancreatic islets into the liver of the patient coupled with a less intensive immunosuppressive treatment (the Edmonton protocol). Expanded clinical trials of this procedure are currently underway. Scientists are also evaluating methods of slowing the original autoimmune destruction of pancreatic beta cells that produces the disease in the first place.

Whole pancreas and islet cell transplants ameliorate Type- 1 diabetes, but there is nowhere near enough of these materials to treat all in need. To overcome this shortage, people hope that human stem cells can be induced—at will and in bulk— to differentiate in vitro into functional pancreatic beta cells, available for transplantation. Of course, it would still be crucial to prevent immunological destruction of the newly transplanted stem cell-derived beta cells.

3. Stem Cell Therapy for Type-1 Diabetes?

Initial exp*eriments in mice suggested that insulinpro*ducing cells could be obtained from mouse embryonic stem cells following in vitro differentiation. Can this approach be extended to human stem cells? A number of attempts have been made, with promising initial findings, yet they are not easily evaluated, partly because the criteria for characterizing the cells are not standardized. In a recent paper, Lechner and Habener provided a list of six criteria to define the characteristics of pancreas-derived "beta-like" cells that could be potentially useful in treatment of Type-1 diabetes. We have used those criteria to facilitate assessment of the current state of progress toward development of functional "beta-like" cells that might eventually be tested in Type-1 diabetes patients. Table 1 summarizes and compares the properties of human cell preparations recently produced in research seeking this objective by Abraham et al., Zulewski et al., Assady et al., Zhao et al., and Zalzman et al., and tested in mouse models of human diabetes.

As the results described in Table 1 indicate, cells derived from some human stem cells transplanted into specific strains of mice mimicking major aspects of Type-1 human diabetes were able to reverse high blood glucose concentrations.

Although these results are encouraging, the transplant rejection question remains unanswered because the likely immune rejection of the transplanted human cells was prevented in these experiments by using special strains of immunodeficient mice that lack the capacity to recognize and attack foreign cells.

No tumors were observed in the transplanted mice, but the experiments were terminated after about three months, an insufficient time for much tumor development to occur. Because many Type-1 diabetes patients are children and because a largely effective therapy (insulin injection) is currently available, the introduction of islet cell transplant therapy will need a high degree of certainty that the introduced cells or their derivatives will not become malignant over the course of the patient's life. Stringent tests of the cancer-causing potential of candidate cell preparations will be required, including multiyear studies in animals that live longer than mice or rats. Longterm follow-up of children and adult patients who had received bone marrow transplants many years ago has revealed an increased risk of severe neurologic complications and a variety of types of cancer. C. Therapeutic Applications of Mesenchymal Stem Cells (MSCs) Before stem cell based therapies are used to treat human diseases, they will have to gain approval through the Food and Drug Administration (FDA) regulatory process.

The first step in this process is filing an Investigational New Drug (IND) application. As of July 2003, four IND applications have been filed for clinical applications of mesenchymal stem cells. The disease indications include: (1) providing MSC support for peripheral blood stem cell transplantation in cancer treatment, (2) providing MSC support for cord blood transplantation in cancer treatment, (3) using MSCs to stimulate regeneration of cardiac tissue after acute myocardial infarction (heart attack), and (4) using MSCs to stimulate regeneration of cardiac tissue in cases of congestive heart failure.

The first two applications are currently in Phase II of the regulatory process, with pivotal Phase III trials scheduled to begin in 2004. D. Evaluating the Different Types of Stem Cells A major unresolved issue at present involves the therapeutic potential of human adult stem cells compared with embryonic stem cells. The answer may well be different for different diseases and for patients of different ages.

For example, in treating an elderly patient with Parkinson's Disease, the use of adult stem cells may be appropriate even if these cells may have a more limited number of cell divisions remaining. On the other hand, treating a child with Type I Diabetes, one may want to use embryonic stem cells because of their potentially greater longevity, or other factors. The only valid way to resolve these questions is by instituting rigorous therapeutic trials which test the efficacy of the different types of stem cells in treating a variety of different diseases to determine their comparative efficacy. Clearly, such trials would be a long-term endeavor, since it would take years to obtain answers to these very critical questions.

Private Sector Activity

In the United States, much of the basic research on animal stem cells and human adult stem cells has been publicly funded. Yet before *2001,* research in the U.S., using human ESCs could only be done in the private sector (the locus also of much research on animal and human adult stem cells). The current state of knowledge about human ESCs (and also about human MSCs) reflects pioneering and on-going stem cell research funded by the private sector in the U.S.

For example, the work that led to the 1998 reports of the first isolation of both ESCs and EGCs, was funded by Geron Corporation. Embryonic and adult stem cell research is today vigorously pursued by many companies and supported by several private philanthropic foundations, and the results of some of this research have been published in peer-reviewed journals.

Private sector organizations have pursued and been awarded patents on the stem cells themselves and methods for producing and using them to treat disease. As noted above, at least one company (Osiris Therapeutics) has protocols under review at the FDA for clinical trials with MSCs. It seems likely that private sector companies will continue to play large roles in the future development of stem cell based therapies.

Preliminary Conclusions

While it might be argued that it is too soon to attempt to draw any conclusions about the state of a field that is changing as rapidly as stem cell research, we draw the following preliminary conclusions regarding the current state of the field.

Human stem cells can be reproducibly isolated from a variety of embryonic, fetal, and adult tissue sources. Some human stem cell preparations (for example, human ESCs, EGCs, MSCs, and MAPCs) can be reproducibly expanded to substantially larger cell numbers in vitro, the cells can be stored frozen and recovered, and they can be characterized and compared by a variety of techniques. These cells are receiving a large share of the attention regarding possible future (nonhematopoietic) stem cell transplantation therapies.

Preparations of ESCs, EGCs, MSCs, and MAPCs can be induced to differentiate in vitro into a variety of cells with properties similar to those found in differentiated tissues.

Research using these human stem cell preparations holds promise for: (a) increased understanding of the basic molecular process underlying cell differentiation, (b) increased understanding of the early stages of genetic diseases (and possibly cancer), and (c) future cell transplantation therapies for human diseases.

The case study of developing stem cell-based therapies for Type-1 diabetes illustrates that, although insulin-producing cells have been derived from human stem cell preparations, we could still have a long way to go before stem cell-based therapies can be developed and made available for this disease. This appears to be true irrespective of whether one starts from human embryonic stem cells or from human adult stem cells. The transplant rejection problem remains a major obstacle, but only one among many.

Human mesenchymal stem cells are currently being evaluated in pre-clinical studies and clinical trials for several specific human diseases.

Much basic and applied research remains to be done if human stem cells are to achieve their promise in regenerative medicine. This research is expensive and technically challenging, and requires scientists willing to take a long perspective in order to discover, through painstaking research, which combinations of techniques could turn out to be successful. Strong financial support, public and private, will be indispensable to achieving success.

Contributors to this journal include Under Secretary of State Alan Larson, Under Secretary of Agriculture J.B. Penn, Deputy Food and Drug Administration Commissioner Lester Crawford, and Ambassador Tony Hall, U.S. Representative to the U.N. Agencies for Food and Agriculture, who address a broad range of topics from the basic science of biotechnology to food safety and labeling issues. Their articles are complemented by essays from an internationally respected group of researchers and academics, a State Department fact sheet on the Cartagena Biosafety Protocol and additional resource information.

Millions of farmers worldwide now plant biotech crops, agency says:

> The U.S. Agency for International Development (USAID) has issued a series of six fact sheets outlining the key issues related to products derived from modern biotechnology. The first fact sheet defines agricultural biotechnology.

For about 10,000 years, farmers have been improving wild plants and animals through the selection and breeding of desirable characteristics. This breeding has resulted in the domesticated plants and animals that are commonly used in crop and livestock agriculture. In the twentieth century, breeding became more sophisticated, as the traits that breeders select for include increased yield, disease and pest resistance, drought resistance and enhanced flavor. Traits are passed from one generation to the next through genes, which are made of DNA. All living things — including the fruits, vegetables and meat that we eat — contain genes that tell cells how to function. Recently, scientists have learned enough to begin to identify and work with the genes (DNA) that are responsible for traits.

Agricultural biotechnology is a collection of scientific techniques used to improve plants, animals and microorganisms. Based on an understanding of DNA, scientists have developed solutions to increase agricultural productivity. Starting from the ability to identify genes that may confer advantages on certain crops, and the ability to work with such characteristics very precisely, biotechnology enhances breeders' ability to make improvements in crops and livestock. Biotechnology enables improvements that are not possible with traditional crossing of related species alone.

Genetic engineering: Scientists have learned how to move genes from one organism to another. This has been called genetic modification (GM), genetic engineering (GE) or genetic improvement (GI). Regardless of the name, the process allows the transfer of useful characteristics (such as resistance to a disease) into a plant, animal or microorganism by inserting genes (DNA) from another organism. Virtually ail crops improved with transferred DNA (often called GM crops or GMOs) to date have been developed to aid farmers to increase productivity by reducing crop damage from weeds, diseases or insects.

Molecular markers: Traditional breeding involves

selection of individual plants or animals based on visible or measurable traits. By examining the DNA of an organism, scientists can use molecular markers to select plants or animals that possess a desirable gene, even in the absence of a visible trait. Thus, breeding is more precise and efficient. For example, the International Institute of Tropical Agriculture has used molecular markers to obtain cowpea resistant to bruchid (a beetle), disease-resistant white yam and cassava resistant to Cassava Mosaic Disease, among others. Another use of molecular markers is to identify undesirable genes that can be eliminated in future generations.

Molecular diagnostics: Molecular diagnostics are methods to detect genes or gene products that are very precise and specific. Molecular diagnostics are used in agriculture to more accurately diagnose crop/livestock diseases.

Vaccines: Biotechnology-derived vaccines are used in livestock and humans. They may be cheaper, better and/or safer than traditional vaccines. They are also stable at room temperature, and do not need refrigerated storage; this is an important advantage for smallholders in tropical countries. Some are new vaccines, which offer protection for the first time against some infectious illnesses. For example, in the Philippines, biotechnology has been used to develop an improved vaccine to protect cattle and water buffalo against hemorrhagic septicemia, a leading cause of death for both species.

Tissue culture: Tissue culture is the regeneration of plants in the laboratory from disease-free plant parts. This technique allows for the reproduction of disease-free planting material for crops. Examples of crops produced using tissue culture include citrus, pineapples, avocados, mangoes, bananas, coffee and papaya.

The first food product of biotechnology (an enzyme used in cheese production and a yeast used for baking) appeared on the market in 1990. Since 1995, farmers have been growing GE crops. In 2003, 7 million farmers in 18 countries — more than 85 percent of them resource-poor farmers in the developing world — were planting biotech crops. Almost one third of the global biotech crop area was grown in developing countries.

A safe and sufficient food supply, grown in an environmentally responsible fashion is essential for humanity. Like any technology, agricultural biotechnology will have economic and social impacts. Since their introduction, crops improved using biotechnology have been used safely, with benefits such as the reduction of pesticide use. Agricultural biotechnology is only one factor among many influencing the health and welfare of farmers and other citizens in the developing world. As biotechnology continues to evolve, factual and open public discourse is vital to define the rote it should play in Society.

REFERENCES

Carroll, J.S.: 2001, 'Social Science Research Methods for Assessing Societal Implications of Nanotechnology', in: M.C. Roco and W.S. Bainbridge (eds.), *Societal Implications of Nanoscience and Nanotechnology*, Dordrecht: Kluwer, pp. 188-192.

Chen, C-Y., S.L. Burkett, H.-X. Li, and M.E. Davis. 1993. *Microporous Mater.* 2:27.

Chianelli, R.R., M. Daage, and M.J. Ledoux. 1994. *Advances in Catalysis* 40:177.

Dresselhaus, M., and G. Dresselhaus. 1995. *Ann. Rev. Mat. Sci.* 25:487.

Dresselhaus, M.S. 1998. Carbon-based nanostructures. In *R&D status and trends*, ed. Siegel et al.

Dresselhaus, M.S., G. Dresselhaus, and P. Eklund. 1996. *Science of fullerenes and carbon nanotubes*. San Diego: Academic Press.

Drexler, K.E.: 2004, 'Nanotechnology: From Feynman to Funding', *Bulletin of Science, Technology & Society*, 24, 21-27.

Dupuy, J.-P.: 2004, 'Complexity and Uncertainty: A Prudential Approach To Nanotechnology', in: European Commission (Community Health and Consumer Protection): *Nanotechnologies: A Preliminary Risk Analysis on the Basis of a Workshop, Brussels, 1-2 March 2004*, pp. 71-93 (www.europa.eu.int/comm/health/ ph_risk/documents/ ev_20040301_en.pdf).

Dupuy, J.-P.: 2004, 'Pour une évaluation normative du programme nanotechnologique', *Annales des Mines*, (February 2004), 27-32.

Einsiedel, E.F.; Goldenberg, L.: 2004, 'Dwarfing the Social? Nanotechnology Lessons from the Biotechnology Front', *Bulletin of Science, Technology & Society*, 24, 28-33.

Etzkowitz, H., 2001, 'Nano-Science and Society: Finding a Social Basis for Science Policy', in: M.C. Roco and W.S. Bainbridge (eds.), *Societal Implications of Nanoscience and Nanotechnology*, Dordrecht: Kluwer, pp. 121-128.

European Commission: 2004, *Converging Technologies: Shaping the Future of European Societies (Report of the High Level Expert Group "Foresighting the New Technology Wave")*, Brussels: European Commission Research.

F. Buot, "Mesoscopic Physics and Nanoelectronics: Nanoscience and Nanotechnology," *Physics Reports*, pp.73-174, 1993.

F. Capasso and S. Datta, "Quantum Electron Devices," *Physics Today*, pp.74-82, February 1990.

Feynman, R., "There's Plenty of Room at the Bottom: An invitation to Enter a New Field of Physics," Talk at the Annual Meeting of the American Physical Society, 29 December 1959. Reprinted in Appendix B of Crandall and Lewis. See citation above.

Fielder, F.A.; Reynolds, G.H.: 1994, 'Legal Problems of Nanotechnology: An Overview', *Southern California Interdisciplinary Law Journal*, 3, 593-629.

Fleischer, Th.; Decker, M. & Fiedeler, U. (eds.): 2004, *Große Aufmerksamkeit für kleine Welten ' Nanotechnologie und ihre Folgen*, special issue of *Technikfolgenabschätzung ' Theorie und Praxis*, 13 (2), 5-85 [www.itas.fzk.de/tatup/042/inhalt.htm].

Fogelberg, H. & Glimell, H.: 2003, 'Molecular Matters: In Search of the Real Stuff', in: H. Fogelberg & H. Glimell, *Bringing Visibility to the Invisible: Towards A Social Understanding of Nanotechnology*, Göteborg: Göteborg University, pp. 5-32 [www.sts.gu.se/publications/STS_report_6.pdf].

Glimell, H.: 2001, 'Dynamics of the Emerging Field of Nanoscience', in: M.C. Roco and W.S. Bainbridge (eds.), *Societal Implications of Nanoscience and Nanotechnology*, Dordrecht: Kluwer, pp. 156-160.

Glimell, H.: 2003, 'A Nano Narrative: The Mircopolitics of a New Generic Enabling Technology', in: H. Fogelberg & H. Glimell, *Bringing Visibility to the Invisible: Towards A Social Understanding of Nanotechnology*, Göteborg: Göteborg University, pp. 55-77 [www.sts.gu.se/publications/STS_report_6.pdf].

Glimell, H.: 2003, 'Dynamics of the Emerging Field of Nanoscience', in: H. Fogelberg & H. Glimell, *Bringing Visibility to the Invisible: Towards A Social Understanding of Nanotechnology*, Göteborg: Göteborg University, pp. 79-85 [www.sts.gu.se/publications/STS_report_6.pdf].

Glimell, H.: 2004, 'Grand Visions and Lilliput Politics: Staging the Exploration of the 'Endless Frontier'', in: D. Baird, A. Nordmann & J. Schummer (eds.), *Discovering the Nanoscale*, Amsterdam: IOS Press, pp. 231-246.

Goldhaber-Gordon, D., Montemerlo, M.S., Love, J.C., Opiteck, G.J., and Ellenbogen, J.C., "Overview of Nanoelectronic Devices," submitted to the *Proceedings of the IEEE*, February 1997. For more information, please send e-mail to nanotech@mitre.org.

Gorman, M.E.; Groves, J.F. & Shrager, J.: 2004, 'Societal Dimensions of Nanotechnology as a Trading Zone: Results from a Pilot Project', in: D. Baird, A. Nordmann & J. Schummer (eds.), *Discovering the Nanoscale*, Amsterdam: IOS Press, pp. 63-73.

Grinbaum, A. & Dupuy, J.-P.: 2004, 'Living with Uncertainty: Toward the Ongoing Normative Assessment of Nanotechnology', *Techne: Research in Philosophy and Technology*, 8(3) (forthcoming).

Grinbaum, A.: 2004, 'La condition de l'homme moderne et les nanotechnologies', in: G. Nivat (ed.), *Les limites de l'humain. 39èmes Rencontres Internationales de Genève*, L'Age d'Homme, Genève, p. 141.

Grunwald, A.: 2004, 'Ethische Aspekte der Nanotechnologie. Eine Felderkundung', *Technikfolgenabschätzung ' Theorie und Praxis*, 13 (2), 71-78.

Gupta, V.K. & Pangannaya, N.B.: 2000, 'Carbon nanotubes: bibliometric analysis of patents', *World Patent Information*, 22, 185-189.

Hansson, S.O.: 2004, 'Great Uncertainty about Small Things', *Techne: Research in Philosophy and Technology*, 8(3) (forthcoming).

Haruta, M. 1997. *Catalyst surveys of Japan* 1:61 and references therein.

Hennig, J.: 2004, 'Changes in the Design of Scanning Tunneling Microscopic Images from 1980 to 1990', *Techne: Research in Philosophy and Technology*, 8(3) (forthcoming).

Hessenbruch, A.: 2004, 'Nanotechnology and the Negotiation of Novelty', in: D. Baird, A. Nordmann & J. Schummer (eds.), *Discovering the Nanoscale*, Amsterdam: IOS Press, pp. 135-144.

Hullmann, A. & Meyer, M.: 2003, 'Publications and Patents in Nanotechnology. An overview of previous studies and the state of the art', *Scientometrics*, 58 (3) 507-527.

Huo, Q., D.I. Margolese, U. Ciesla, P. Feng, T.E. Gier, P. Sieger, R. Leon, P.M. Petroff, F. Schuth, and G.D. Stucky. 1994. *Nature* 368:317.

Jena, P., S.N. Khanna, and B.K. Rao. 1996. In *Science and technology of atomically engineered materials*, ed. P. Jena. River Edge, NJ: World Scientific.

Johansson, M.: 2003, 'Plenty of room at the bottom: Towards an anthropology of nanoscience', *Anthropology Today*, 19 (No. 6), 3-6.

Johnson, A.: 2004, 'The End of Pure Science: Science Policy from Bayh-Dole to the NNI', in: D. Baird, A. Nordmann & J. Schummer (eds.), *Discovering the Nanoscale*, Amsterdam: IOS Press, pp. 217-230.

Jones, C.W., and W.J. Koros. 1994. *Carbon* 32:1419.

Jounet, C., W.K. Maser, P. Bernier, A. Loiseau, M. Lamy de la Chapelle, S. Lefrant, P. Deniard, R. Lee, and J.E. Fischer. 1997. *Nature* 388:756.

K.E. Drexler, *Nanosystems: Molecular Machinery, Manufacturing, and Computation,* Wiley, New York, 1992.

Karger, J., and D.M. Ruthven. 1992. *Diffusion in zeolites*. New York: J. Wiley.

Keiper, A.: 2003, 'The Nanotechnology Revolution', *The New Atlantis*, 2, 17-34.

Khushf, G.: 2004, 'A Hierarchical Architecture for Nano-scale Science and Technology: Taking Stock of the Claims About Science Made By Advocates of NBIC Convergence', in: D. Baird, A. Nordmann & J. Schummer (eds.), *Discovering the Nanoscale*, Amsterdam: IOS Press, pp. 21-33.

Khushf, G.: 2004, 'Systems Theory and the Ethics of Human Enhancement: A Framework for NBIC Convergence', *Annals of the New York Academy of Sciences*, 1013, 124-149.

Krätschmer, W., L.D. Lamb, K. Fostiropoulos, and D.R. Huffman. 1990. *Nature* 347:354.

Kresge, C.T., M.E. Leonowicz, W.J. Roth, J.C. Vartuli, and J.S. Beck. 1992. *Nature* 359:710.

Kupperman, A., S. Nadimi, S. Oliver, G. Ozin, J. Garcés, and M. Olken. 1993. *Nature* 365:239.

Kuusi, O.; Meyer, M.: 2002, 'Technological generalizations and leitbilder ' the anticipation of technological opportunities', *Technological Forecasting & Social Change*, 69, 625-639.

Landon, B.: 2004, 'Less is More: Much Less is Much More: The Insistent Allure of Nanotechnology Narratives in Science Fiction', in: N.K. Hayles (ed.), *Nanoculture: Implications of the New Technoscience*, Bristol, UK: Intellect Books, pp. 131-146.

Laszlo, P.: 2004, 'Is There Life After Partington?', *Hyle: International Journal for Philosophy of Chemistry*, 10(2), 169-178.

Lenhard, J.: 2004, 'Nanoscience and the Janus-Faced Character of Simulations', in: D. Baird, A. Nordmann & J. Schummer (eds.), *Discovering the Nanoscale*, Amsterdam: IOS Press, pp. 93-100.

Lent, C.S., Tougaw, P.D., Porod, W., Bernstein, G.H., "Quantum Cellular Automata," *Nanotechnology*, Vol. 4, p. 49, 1993.

Lewak, S.E.: 2004, 'What's the Buzz? Tell Me What's A-Happening: Wonder, Nanotechnology, and Alice's Adventures in Wonderland', in: N.K. Hayles (ed.), *Nanoculture: Implications of the New Technoscience*, Bristol, UK: Intellect Books, pp. 201-.

Lide, D.R., ed. 1993-1994. *CRC Handbook of Chemistry and Physics*, 74th ed.

Lin-Easton, P.C.: 2001, 'It's Time for Environmentalists to Think Small ' Real Small: A Call for the Involvement of Environmental Lawyers in Developing Precautionary Policies Molecular Nanotechnology', *Georgetown International Law Review*, 14, 106-134.

López, J.: 2004, 'Bridging the Gaps: Science Fiction in Nanotechnology', *Hyle: International Journal for Philosophy of Chemistry*, 10(2), 129-152.

Lösch, A.: 2004, 'Nanomedicine and Space: Discursive Orders of Mediating Innovations', in: D. Baird, A. Nordmann & J. Schummer (eds.), *Discovering the Nanoscale*, Amsterdam: IOS Press, pp. 193-202.

Marshall, K.: 2004, 'Atomizing the Risk Technology', in: N.K. Hayles (ed.), *Nanoculture: Implications of the New Technoscience*, Bristol, UK: Intellect Books, pp. 147-160.

Martin, T.P., N. Malinowski, U. Zimmerman, U. Naher, and H. Schaber. 1993. *J. Chem. Phys.* 99:4210.

Martin, T.P., U. Naher, H. Schaber, U. Zimmerman. 1993. *Phys. Rev. Lett.* 70:3079.

Mayer, S.: 2002, 'From genetic modification to nanotechnology: the dangers of 'sound science'', in: T. Gilland (ed.), *Science: Can We Trust the Experts?*, London: Hodder and Stoughton, pp. 1-15.

Mehta, M.D.: 2002, 'Nanoscience and Nanotechnology: Assessing the Nature of Innovation in These Fields', *Bulletin of Science, Technology & Society*, 22(4), 269-273.

Mehta, M.D.: 2004, 'From Biotechnology to Nanotechnology: What Can We Learn from Earlier Technologies?, *Bulletin of Science, Technology & Society*, 24, 34-39.

Meyer, M. & Kuusi, O.: 2004, 'Nanotechnology: Generalizations in an Interdisciplinary Field of Science and Technology', *Hyle: International Journal for Philosophy of Chemistry*, 10(2), 153-168.

Meyer, M.: 2000a, 'Does science push technology? Patents citing scientific literature', *Research Policy*, 29, 409-434.

Meyer, M.: 2000b, 'Patent citations in a novel field of technology: What can they tell about interactions of emerging communities of science and technology?', *Scientometrics*, 48, 151-178.

Meyer, M.: 2001a, 'Patent citations in a novel field of technology: An exploration of nano-science and nano-technology', *Scientometrics*, 51, 163-183.

Meyer, M.: 2001b, 'Socio-economic Research on Nanoscale Science and Technology: A European Overview and Illustration', in: M.C. Roco and W.S. Bainbridge (eds.), *Societal Implications of Nanoscience and Nanotechnology*, Dordrecht: Kluwer, pp. 217-241.

Milburn, C.: 2002, 'Nanotechnology in the age of post-human engineering: science fiction as science', *Configurations*, 10, 261-295 [reprinted in: N.K. Hayles (ed.), *Nanoculture: Implications of the New Technoscience*, Bristol, UK: Intellect Books, 2004, pp. 109-130].

Milburn, C.: 2004, 'Nano/Splatter: Disintegrating the Postbiological Body', *New Literary History*, 35 (in print).

Mnyusiwalla, A.; Abdallah, S.D.; Singer, P.A.: 2003, 'Mind the gap: science and ethics in nanotechnology', *Nanotechnology*, 14, R9-R13.

Mody, C.C.M.: 2004, 'How Probe Microscopists Became Nanotechnologists', in: D. Baird, A. Nordmann & J. Schummer (eds.), *Discovering the Nanoscale*, Amsterdam: IOS Press, pp. 119-133.

Mody, C.C.M.: 2004, 'Instruments in Training: The Growth of American Probe Microscopy in the 1980s', in: D. Kaiser (ed.), *Pedagogy and the Practice of Science: Producing Physical Scientists, 1800-2000*, Cambridge, MA: MIT Press (forthcoming).

Mody, C.C.M.: 2004, 'Small, but Determined: Technological Determinism in Nanoscience', *Hyle: International Journal for Philosophy of Chemistry*, 10(2), 99-128.

Mody, C.C.M.: 2004, *Crafting the Tools of Knowledge: The Invention, Spread, and Commercialization of Probe Microscopy, 1960-2000*, Ph.D. dissertation, Cornell University.

Montemerlo, M.S., Love, J.C., Opiteck, G.J., Goldhaber, D. J., and Ellenbogen, J.C., "Technologies and Designs for Electronic Nanocomputers," MITRE Technical Report 96W0000044, The MITRE Corporation, McLean, VA, July 1996. For more information, send e-mail to nanotech@mitre.org.

Moor, J.H. & Weckert, J.: 2004, 'Nanoethics: Assessing the Nanoscale From an Ethical Point of View', in: D. Baird, A. Nordmann & J. Schummer (eds.), *Discovering the Nanoscale*, Amsterdam: IOS Press, pp. 301-310.

Munn Sanchez, E.: 2004, 'The Expert's Role in Nanoscience and Technology', in: D. Baird, A. Nordmann & J. Schummer (eds.), *Discovering the Nanoscale*, Amsterdam: IOS Press, pp. 257-266.

Nordmann, A.: 2003, 'Shaping the World Atom by Atom: Eine nanowissenschaftliche WeltBildanalyse', in: A. Grunwald (ed.), *Technikgestaltung zwischen Wunsch und Wirklichkeit*, Berlin: Springer, pp. 191-199.

Nordmann, A.: 2004, 'Molecular Disjunctions: Staking Claims at the Nanoscale', in: D. Baird, A. Nordmann & J. Schummer (eds.), *Discovering the Nanoscale*, Amsterdam: IOS Press, pp. 51-62.

Nordmann, A.: 2004, 'Nanotechnology: Convergence and Integration', Presentation at *EuroNanoForum, Trieste, December 10, 2003* (Proceedings in preparation).

Nordmann, A.: 2004, 'Nanotechnology's WorldView: New Space for Old Cosmologies', *IEEE Technology and Society Magazine*, 23 (forthcoming).

Nordmann, A.: 2004, 'Social Imagination for Nanotechnology', in: European Commission (Community Health and Consumer Protection): *Nanotechnologies: A Preliminary Risk Analysis on the Basis of a Workshop, Brussels, 1-2 March 2004*, pp. 111-113 [www.europa.eu.int/comm/health/ph_risk/documents/ev_20040301_en.pdf].

Nordmann, A.: 2004, 'Was ist TechnoWissenschaft? ' Zum Wandel der Wissenschaftskultur am Beispiel von Nanoforschung und

Bionik', in: T. Rossmann & C. Tropea (eds.), *Bionik ' Neue Forschungsergebnisse aus Natur-, Ingenieur- und Geisteswissenschaften*, Berlin: Springer, 2004 (forthcoming)

Paschen, H.; Coenen, C.; Fleischer, T.; Grünwald, R.; Oertel, D. & Revermann, C.: 2004, *Nanotechnologie: Forschung, Entwicklung, Anwendung*, Berlin: Springer (*Nanotechnologie, TAB-Arbeitsbericht 92*, Berlin: Büro für Technikfolgen-Abschätzung beim Deutschen Bundestag).

Pitt, J.C.: 2004, 'The Epistemology of the Very Small', in: D. Baird, A. Nordmann & J. Schummer (eds.), *Discovering the Nanoscale*, Amsterdam: IOS Press, pp. 157-163.

Pressman, J.: 2004, 'Nano Narrative: A Parable from Electronic Literature', in: N.K. Hayles (ed.), *Nanoculture: Implications of the New Technoscience*, Bristol, UK: Intellect Books, pp. 191-200.

Prigogine, I., and S. Rice. 1988. *Advances in chemical physics*, Vol. 70, Parts 1 & 2. New York: J. Wiley.

R. T. Bate, "Nanoelectronics," *Nanotechnology*, Vol. 1, pp. 1-7, 1990.

Rademann, K., B. Kaiser, U. Even, F. Hensel. 1987. *Phys. Rev. Lett.* 59:2319.

Rao, C.N.R., B.C. Satishkumar, and A. Govindaraj. 1997. *Chem. Commun.* 1581.

Rao, M.B., and S. Sircar. 1993. *Gas Separation and Purification* 7:279.

Reed, M.A., "Quantum Dots," *Scientific American*, January 1993, pp. 118-123.

Roberts, J.A.: 2004, 'Deciding the Future of Nanotechnologies: Legal Perspectives on Issues of Democracy and Technology', in: D. Baird, A. Nordmann & J. Schummer (eds.), *Discovering the Nanoscale*, Amsterdam: IOS Press, pp. 247-255.

Robinson, C.: 2004, 'Images in NanoScience/Technology', in: D. Baird, A. Nordmann & J. Schummer (eds.), *Discovering the Nanoscale*, Amsterdam: IOS Press, pp. 165-169.

Robison, W.L.: 2004, 'Nano-Ethics', in: D. Baird, A. Nordmann & J. Schummer (eds.), *Discovering the Nanoscale*, Amsterdam: IOS Press, pp. 285-300.

Roco, M.C. & Bainbridge, W.S. (eds.): 2001, *Societal implications of nanoscience and nanotechnology*, (Proceedings of a workshop organized by the National Science Foundation, September 28-29, 2000), Kluwer: Dordrecht [available online at http://itri.loyola.edu/nano/societalimpact/nanosi.pdf]

Roco, M.C. & Tomellini, R. (eds.): 2002, *Nanotechnology: Revolutionary Opportunities and Societal Implications, Workshop, Lecce (Italy), 31 January - 1 February 2002*, Luxemburg: Office for Official Publications of the European Communities, [200 pp.]

Roco, M.C.; Bainbridge, W.S. (eds.): 2002, *Converging Technologies for Improving Human Performance: Nanotechnology, Biotechnology, Information Technology and the Cognitive Science*, Arlington, VA: National Science Foundation.

3

Biotech Information Mapping

Imagine you're in a car driving down the highway to visit an old friend who has just moved to Los Angeles. Your favorite tunes are playing on the radio, and you haven't a care in the world. You stop to check your maps and realize that all you have are interstate highway maps—not a single street map of the area. How will you ever find your friend's house? It's going to be difficult, but eventually, you may stumble across the right house.

> Genetic maps serve to guide a scientist toward a gene, just like an interstate map guides a driver from city to city. Physical maps are more similar to street maps and allow a scientist to more easily home in on a gene's location.

This scenario is similar to the situation facing scientists searching for a specific gene somewhere within the vast human genome. They have available to them two broad categories of maps: *genetic maps* and *physical maps*. Both genetic and physical maps provide the likely order of items along a chromosome. However, a genetic map, like an interstate highway map, provides an indirect estimate of the distance between two items and is limited to ordering certain items. One could say that genetic maps serve to guide a scientist toward a gene, just like an interstate map guides a driver from city to city. On the other hand, physical maps mark an estimate of the true distance, in measurements called *base pairs*, between items of interest. To continue our analogy, physical maps would then be similar to street maps, where the distance between two sites of interest may be defined more precisely in terms of city blocks or street addresses. Physical maps, therefore, allow a scientist to more easily home in on the location of a gene. An appreciation of how each of these maps is constructed may be helpful in understanding how scientists use these maps to traverse that genetic highway commonly referred to as the "human genome".

PART I: GENETIC MAPS

Types of Landmarks Found on a Genetic Map

Just like interstate maps have cities and towns that serve as landmarks, genetic maps have landmarks known as *genetic markers*, or "markers" for short. The term marker is used very broadly to describe any observable variation that results from an alteration, or mutation, at a single genetic locus. A marker may be used as one landmark on a map if, in most cases, that stretch of DNA is inherited from parent to child according to the standard rules of inheritance. Markers can be within genes that code for a noticeable physical characteristic such as eye color, or a not so noticeable trait such as a disease. *DNA-based reagents* can also serve as markers. These types of markers are found within the non-coding regions of genes and are used to detect unique regions on a chromosome. DNA markers are especially useful for generating genetic maps when there are occasional, predictable mutations that occur during *meiosis*—the formation of gametes such as egg and sperm—that, over many generations, lead to a high degree of variability in the DNA content of the marker from individual to individual.

Genetic maps use landmarks called genetic markers to guide researchers on their gene hunt.

Commonly Used DNA Markers

- **RFLPs**, or **restriction fragment length polymorphisms**, were among the first developed DNA markers. RFLPs are defined by the presence or absence of a specific site, called a restriction site, for a bacterial restriction enzyme. This enzyme breaks apart strands of DNA wherever they contain a certain nucleotide sequence.
- **VNTRs**, or **variable number of tandem repeat polymorphisms**, occur in non-coding regions of DNA. This type of marker is defined by the presence of a nucleotide sequence that is repeated several times. In each case, the number of times a sequence is repeated may vary.
- **Microsatellite polymorphisms** are defined by a variable number of repetitions of a very small number of base pairs within a sequence. Oftentimes, these repeats consist of the nucleotides, or bases, cytosine and adenosine. The number of repeats for a given microsatellite may differ between individuals, hence the term **polymorphism**—the existence of different forms within a population.
- **SNPs**, or **single nucleotide polymorphisms**, are individual point mutations, or substitutions of a single nucleotide, that do not change the overall length of the DNA sequence in that region. SNPs occur throughout an individual's genome.

From Linkage Analysis to Genetic Mapping

Early geneticists recognized that genes are located on chromosomes and believed that each individual chromosome was inherited as an intact unit. They hypothesized that if two genes were located on the same chromosome, they were physically linked together and were inherited together. We now know that this is not always the case. Studies conducted around 1910 demonstrated that very few pairs of genes displayed complete linkage. Pairs of genes were either inherited independently or displayed *partial linkage*—that is, they were inherited together sometimes, but not always. It is the behavior of chromosomes during meiosis that determines whether two genes will remain linked.

During *meiosis*—the process whereby gametes (eggs and sperm) are produced— two copies of each chromosome pair become physically close. The chromosome arms can then undergo breakage and exchange segments of DNA, a process referred to as recombination or crossing-over. If recombination occurs, each chromosome found in the gamete will consist of a "mixture" of material from both members of the chromosome pair. Thus, recombination events directly affect the inheritance pattern of those genes involved.

Because one cannot physically see crossover events, it is difficult to determine with any degree of certainty how many crossovers have actually occurred. But, using the phenomenon of *co-segregation of alleles of nearby markers*, researchers can reverse-engineer meiosis and identify markers that lie close to each other. Then, using a statistical technique called *genetic linkage analysis*, researchers can infer a likely crossover pattern, and from that an order of the markers involved. Researchers can also infer an estimate for the probability that a recombination occurs between each pair of markers.

> An **allele** is one of the variant forms of a DNA sequence at a particular locus, or location, on a chromosome. Co-segregation of alleles refers to the movement of each marker during meiosis. If a marker tends to "travel" with the disease status, the markers are said to co-segregate.
>
> **Monotonic** functions are functions that tend to move only in one direction.

If recombination occurs as a random event, then two

markers that are close together should be separated less frequently than two markers that are more distant from one another. The recombination probability between two markers, which can range from 0 to 0.5, increases *monotonically* as the distance between the two markers increases along a chromosome. Therefore, the recombination probability may be used as a surrogate for ordering genetic markers along a chromosome. If you then determine the recombination frequencies for different pairs of markers, you can construct a map of their relative positions on the chromosome.

> Linkage maps can tell you where markers are in relation to each other on the chromosome, but the actual "mileage" between those markers may not be so well defined.

But alas, predicting recombination is not so simple. Although crossovers are random, they are not uniformly distributed across the genome or any chromosome. Some chromosomal regions, called *recombination hotspots*, are more likely to be involved in crossovers than other regions of a chromosome. This means that genetic map distance does not always indicate physical distance between markers. Despite these qualifications, linkage analysis usually correctly deduces marker order, and distance estimates are sufficient to generate genetic maps that can serve as a valuable framework for genome sequencing.

Linkage Studies in Patient Populations: Genetic Maps and Gene Hunting

> In humans, genetic diseases are frequently used as gene markers, with the disease state being one allele and the healthy state the second allele.

In humans, data for calculating recombination frequencies are obtained by examining the genetic makeup of the members of successive generations of existing families, termed *human pedigree analysis*. Linkage studies begin by obtaining blood samples from a group of related individuals. For relatively rare diseases, scientists find a few large families that have many cases of the disease and obtain samples from as many family members as possible. For more common diseases where the pattern of disease inheritance is unclear, scientists will identify a large number of affected families and will take samples from four to thirty close relatives. DNA is then harvested from all of the blood samples and screened for the presence, or *co-inheritance*, of two markers. One marker is usually the gene of interest, generally associated with a physically identifiable characteristic. The other is usually one of the various detectable rearrangements mentioned earlier, such as a microsatellite. A computerized analysis is then performed to determine whether the two markers are linked and approximately how far apart those markers are from one another. In this case, the value of the genetic map is that an inherited disease can be located on the map by following the inheritance of a DNA marker present in affected individuals but absent in unaffected individuals, although the molecular basis of the disease may not yet be understood, nor the gene(s) responsible identified.

Genetic Maps as a Framework for Physical Map Construction

Genetic maps are also used to generate the essential backbone, or scaffold, needed for the creation of more detailed human genome maps. These detailed maps, called *physical maps*, further define the DNA sequence between genetic markers and are essential to the rapid identification of genes.

PART II: PHYSICAL MAPS

Types of Physical Maps and What They Measure

Physical maps can be divided into three general types: *chromosomal* or *cytogenetic maps*, *radiation hybrid (RH) maps*, and *sequence maps*. The different types of maps vary in their degree of *resolution*, that is, the ability to measure the separation of elements that are close together. The higher the resolution, the better the picture.

The lowest-resolution physical map is the chromosomal or *cytogenetic map*, which is based on the distinctive banding patterns observed by light microscopy of stained chromosomes. As with genetic linkage mapping, chromosomal mapping can be used to locate genetic markers defined by traits observable only in whole organisms. Because chromosomal maps are based on estimates of physical distance, they are considered to be physical maps. Yet, the number of base pairs within a band can only be estimated.

RH maps and sequence maps, on the other hand, are more detailed. RH maps are similar to linkage maps in that they show estimates of distance between genetic and physical markers, but that is where the similarity ends. *RH maps* are able to provide more precise information regarding the distance between markers than can a linkage map.

The physical map that provides the most detail is the sequence map. *Sequence maps* show genetic markers, as well as the sequence between the markers, measured in base pairs.

RH Mapping

RH mapping, like linkage mapping, shows an estimated distance between genetic markers. But, rather than relying on natural recombination to separate two markers, scientists use breaks induced by radiation to determine the distance between two markers. In RH mapping, a scientist exposes

DNA to measured doses of radiation, and in doing so, controls the average distance between breaks in a chromosome. By varying the degree of radiation exposure to the DNA, a scientist can induce breaks between two markers that are very close together. The ability to separate closely linked markers allows scientists to produce more detailed maps. RH mapping provides a way to localize almost any genetic marker, as well as other genomic fragments, to a defined map position, and RH maps are extremely useful for ordering markers in regions where highly *polymorphic genetic markers* are scarce.

> **Polymorphic** refers to the existence of two or more forms of the same gene, or genetic marker, with each form being too common in a population to be merely attributable to a new mutation. Polymorphism is a useful genetic marker because it enables researchers to sometimes distinguish which allele was inherited.

Sequence Mapping

Sequence tagged site (STS) mapping is another physical mapping technique. An STS is a short DNA sequence that has been shown to be unique. To qualify as an STS, the exact location and order of the bases of the sequence must be known, and this sequence may occur only once in the chromosome being studied or in the genome as a whole if the DNA fragment set covers the entire genome.

Common Sources of ETS's

- **Expressed sequence tags (ESTs)** are short sequences obtained by analysis of complementary DNA (cDNA) clones. Complementary DNA is prepared by converting mRNA into double-stranded DNA and is thought to represent the sequences of the genes being expressed. An EST can be used as an STS if it comes from a unique gene and not from a member of a gene family in which all of the genes have the same, or similar, sequences.

- **Simple sequence length polymorphisms (SSLPs)** are arrays of repeat sequences that display length variations. SSLPs that are polymorphic and have already been mapped by linkage analysis are particularly valuable because they provide a connection between genetic and physical maps.
- **Random genomic sequences** are obtained by sequencing random pieces of cloned genomic DNA or by examining sequences already deposited in a database.

To map a set of STSs, a collection of overlapping DNA fragments from a chromosome is digested into smaller fragments using *restriction enzymes*, agents that cut up DNA molecules at defined target points. The data from which the map will be derived are then obtained by noting which fragments contain which STSs. To accomplish this, scientists copy the DNA fragments using a process known as "*molecular cloning*". Cloning involves the use of a special technology, called *recombinant DNA technology*, to copy DNA fragments inside a foreign host.

First, the fragments are united with a carrier, also called a vector. After introduction into a suitable host, the DNA fragments can then be reproduced along with the host cell DNA, providing unlimited material for experimental study. An unordered set of cloned DNA fragments is called a *library*.

Next, the *clones*, or copies, are assembled in the order they would be found in the original chromosome by determining which clones contain overlapping DNA fragments. This assembly of overlapping clones is called a *clone contig*. Once the order of the clones in a chromosome is known, the clones are placed in frozen storage, and the information about the order of the clones is stored in a computer, providing a valuable resource that may be used for further studies. These data are then used as the base material for generating a lengthy, continuous DNA sequence, and the STSs serve to anchor the sequence onto a physical map.

The Need to Integrate Physical and Genetic Maps

As with most complex techniques, STS-based mapping has its limitations. In addition to gaps in clone coverage, DNA fragments may become lost or mistakenly mapped to a wrong position. These errors may occur for a variety of reasons. A DNA fragment may break, resulting in an STS that maps to a different position. DNA fragments may also get deleted from a clone during the replication process, resulting in the absence of an STS that should be present. Sometimes a clone composed of DNA fragments from two distinct genomic regions is replicated, leading to DNA segments that are widely separated in the genome being mistakenly mapped to adjacent positions. Lastly, a DNA fragment may become contaminated with host genetic material, once again leading to an STS that will map to the wrong location. To help overcome these problems, as well as to improve overall mapping accuracy, researchers have begun comparing and integrating STS-based physical maps with genetic, RH, and cytogenetic maps. *Cross-referencing* different genomic maps enhances the utility of a given map, confirms STS order, and helps order and orient evolving contigs.

NCBI and Map Integration

Comparing the many available genetic and physical maps can be a time-consuming step, especially when trying to pinpoint the location of a new gene. Without the use of computers and special software designed to align the various maps, matching a sequence to a region of a chromosome that corresponds to the gene location would be very difficult. It would be like trying to compare 20 different interstate and street maps to get from a house in Ukiah, California to a house in Beaver Dam, Wisconsin. You could compare the maps yourself and create your own travel itinerary, but it would probably take a long time. Wouldn't it be easier and faster to have the automobile club create an integrated map for you? That is the goal behind NCBI's Human Genome Map Viewer.

NCBI's Map Viewer: A Tool for Integrating Genetic and Physical Maps

The *NCBI Map Viewer* provides a graphical display of the available human genome sequence data as well as sequence, cytogenetic, genetic linkage, and RH maps. Map Viewer can simultaneously display up to seven maps, selected from a large set of maps, and allows the user access to detailed information for a selected map region. Map Viewer uses a common sequence numbering system to align sequence maps and shared markers as well as gene names to align other maps. You can use NCBI's Map Viewer to search for a gene in a number of genomes, by choosing an organism from the Map Viewer home page.

Map Viewer Getting Started

Need help using the NCBI Map Viewer? Try GETTING STARTED, a quick "how-to guide" on NCBI data mining tools designed for the novice user. GETTING STARTED using Map Viewer provides:

- information on how you can use Map Viewer
- descriptions of the Map Viewer layout
- step-by-step information on using the Map Viewer
- shortcuts for getting to where you need to go.

REFERENCES

Alan Meisel of the University of Pittsburgh School of Law, quoted in Coyle, M., "The Clone Zone," *The National Law Journal*, May 15, 2002.

Asahara T, Takahashi T, Masuda H, Kalka C, Chen D, Iwaguro H, Inai Y, Silver M, Isner JM. VEGF contributes to postnatal neovascularization by mobilizing bone marrow-derived endothelial progenitor cells. EMBO J 1999; 18:3964- 3972.

Asahara, T. et al. Isolation of putative progenitor cells for endothelial angiogenesis. Science 1997; 275:964-967.

Assady, S., *et al.*, "Insulin production by human embryonic stem cells," Diabetes 50: 1691-1697 (2001).

Atkinson, M. A. and Leiter, E. H., "The NOD mouse model of type 1 diabetes: As good as it gets?" Nature Medicine 5: 601-604 (1999).

Baker, K. S., *et al.*, "New Malignancies After Blood or Marrow Stem-Cell Transplantation in Children and Adults: Incidence and Risk Factors," Journal of Clinical Investigation 21: 1352-1358 (2003).

Brivanlou, A. H., *et al.*, "Stem cells. Setting standards for human embryonic stem cells," Science 300: 913-916 (2003).

Bush, G.W., "Stem Cell Science and the Preservation of Life," *The New York Times*, August 12, 2001, p. D13.

Callahan, D., *What Price Better Health: The Hazards of the Research Imperative.*

Carpenter, Janet and Leonard Gianessi, "Herbicide Tolerant Soybeans: Why Growers Are Adopting Roundup Ready Varieties," *AgBioForum*, vol. 2, no. 2, 1999. Available at www.agbioforum.missouri.edu.

Carpenter, M.K., et al., "Characterization and Differentiation of Human Embryonic Stem Cells," Cloning and Stem Cells 5: 79-88 (2003), and Pittenger, *M. F. et al., "Multi*lineage potential of adult human mesenchymal stem cells," Science 284: 143-147 (1999).

Chen, Y., *et al.*, "Embryonic stem cells generated by nuclear transfer of human somatic nuclei into rabbit oocytes," Cell Research 13: 251-264 (2003).

Childress, J., "Federal Policy Toward Embryonic Stem cell research," *American Journal of Bioethics* 2(1): 34 (2002).

Choi K, Kennedy M, Kazarov A, Papadimitriou, Keller G. A common precursor for hematopoietic and endothelial cells. Development 1998;125:725-732.

Chu, K., et *al.*, "Human neural stem cells can migrate, differentiate, and integrate after intravenous transplantation in adult rats with transient forebrain ischemia," Neuroscience Letters 343: 129-133 (2003).

Clark, J., "Squandering Our Technological Future," *The New York Times*, August 30, 2001, p. A19. See also the presentation of Thomas Okarma, President and CEO of Geron Corporation, before the Council on September 4, 2003.

Conference of Catholic Bishops, in a news release ("Catholic Bishops Criticize Bush Policy on Embryo Research," made available by the United States Conference of Catholic Bishops, August 9, 2001).

Daley, G. Q., *et al.*, "Realistic Prospects for Stem Cell Therapeutics," Hematology American Society for Hematology Education *Program: 39*8-418 (2003).

Daley, G., "Cloning and Stem Cells—Handicapping the Political and Scientific Debates," *New England Journal of Medicine* 349(3): 211-212 (2003).

Doerflinger, R., "Ditching Religion and Reality," *American Journal of Bioethics* 2(1): 31 (2002).

Drukker, M., et al., "Characterization of the expression of MHC proteins in human embryonic stem cells," Proceedings of the National Academy of Sciences of th*e United States of America 99: 9864-9869 (2002).*

Elefanty AG, Robb L, Birner R, Begley CG. Hematopoietic-specific genes are not induced during in vitro differentiation of scl-null embryonic stem cells.

Englund, U., *et al.*, "Transplantation of human neural progenitor cells into the neonatal rat brain: extensive migration and differentiation with longdistance axonal projections," Experimental Neurology 173: 1-21 (2002).

Faden, *et al.*, "Public Stem Cell Banks: Considerations of Justice in Stem Cell Research and Therapy" *Hastings Center Report* 33:6 (2003).

Faraci, M., *et al.*, "Severe neurologic complications after hematopoietic stem cell transplantation in children," Neurology 59: 1895-1904 (2002).

Folkman, J. Therapeutic angiogenesis in ischemic limbs. Circulation 1998; 97:108- 110.

Gianessi, Leonard P. and Janet E. Carpenter, "Agricultural Biotechnology: Insect Control Benefits," National Center for Food and Agricultural Policy, 1999. Available on Biotechnology Industry Organization at www.bio.org.

Hogan, P., et al., "Economic Costs of Diabetes in the US in 2002," Diabetes Care 26: 917-932 (2003).

Hori, Y., et al., "Growth inhibitors promote differentiation of insulin-producing tissue from embryonic stem cells," Proceedings of the National Academy of Sciences of the *United States of America 99: 16105-16110 (2002).*

Ivanova, N. B., *et al.,* "A Stem Cell Molecular Signature," Science 298: 601-604 (2002).

Jackson KA, Majka SM, Wang H, Pocius J, Hartley CJ, Majesky MW, Entman ML, Michael LH, Hirschi KK, Goodell MA (2001) Regeneration of ischemic cardiac muscle and vascular endothelium by adult stem cells. J Clin Invest 107:1395-1402.

Jaffredo T, Gautier R, Eichmann A, Dieterlen-Lievre F. Intraaortic hemopoietic cells are derived from endothelial cells during ontogeny. Development 1998;125:4575-4583.

James, Clive, "Global Review of Commercialized Transgenic Crops: 1998," *ISAAA*, No. 8, 1998.

Kahn, J., "Missing the mark on stem cells," *Bioethics Examiner* 5:3, Fall 2001, p. 1.

Kalka, C. et al. Transplantation of ex vivo expanded endothelial progenitor cells for therapeutic neovascularization. Proc. Natl. Acad. Sci. USA 2000; 97:3422-3427.

Kass, L., "The Meaning of Life - In the Laboratory," *The Public Interest*, Winter 2002. Also see the Council's July 2002 report *Human Cloning and Human Dignity: An Ethical Inquiry,* Chapter 6.

Kawano H, Do YS, Kawano Y, Starnes V, Barr M, Law RE, Hsueh WA. Angiotensin II has multiple profibrotic effects in human cardiac fibroblasts. Circulation 2000;101:1130-1137.

Kennedy M, Firpo M, Choi K, Wall C, Robertson S, Kabrun N, Keller

G. A common precursor for primitive erythropoiesis and definitive haematopoiesis.

Kennedy, D., "Stem Cells: Still Here, Still Waiting," *Science* 300: 865 (2003).

Kerr, D. A., et al., "*Hum*an Embryonic Germ Cell Derivatives Facilitate Motor Recovery of Rats with Diffuse Motor Neuron Injury," The Journal of Neuroscience 23: 5131-5140 (2003).

Kerr, D. A., *et al.,* "Human Embryonic Germ Cell Derivatives Facilitate Motor Recovery of Rats with Diffuse Motor Neuron Injury," The Journal of Neuroscience 23: 5131-5140 (2003).

Labastie M-C, Cortes F, Romeo P-H, Dulac C, Peault B. Molecular identity of hematopoietic precursor cells emerging in the human embryo. Blood 1998;92:3624-3635.

Latham, S., "Ethics and Politics," *American Journal of Bioethics* 2(1): 46 (2002).

Lechner, A. and Habener, J.F., "Stem/progenitor cells derived from adult tissues: potential for the treatment of diabetes mellitus," American Journal of Physiology—Endocrinology and M*etabolism* 284: E259-266 (2003).

Lewis, C. S. The Abolition of Man. New York: The Macmillan Company, 1947.

Liker, M., et al., "Human neural stem cell transplantation in the MPTP-lesioned mouse," Brain Research 971: 168-177 (2003).

Liu, Z., and Marti*n, L. J.,* "Olfactory bulb core is a rich source of neural progenitor and stem cells in adult rodent and human," Journal of Comparative Neurology 459: 368-391 (2003).

Liu, Z., and Marti*n,* L.J., "Olfactory bulb core is a rich source of neural progenitor and stem cells in adult rodent and human," Journal of Comparative Neurology 459: 368-391 (2003).

Los Angeles, CA.: University of California Press, 2003; also see Daniel Callahan's presentation before the Council on July 24, 2003, available on the Council's website at www.bioethics.gov.

Lumelsky, N., *et al.,* "Differentiation of embryonic stem cells to insulin-secreting structures similar to pancreatic islets," Science 292: 1389-1394 (2001).

Macklin, Ph.D., Ruth. "Ethics, Politics, and Human Embryo Stem

Cell Research." Women's Health Issues, vol. 10, no. 3 (May/June 2000): 111-15.

Magnus, David, Arthur Caplan, and Glenn McGee, ed. In Who Owns Life? Amherst, New York: Prometheus Books, 2002.

Majumdar, M. K., et al., "Characterization and functionality of cell surface molecules on human mesenchymal stem cells," Journal of Biomedical Science 10: 228-241 (2003).

Marquis, D., "Stem cell research: The Failure of Bioethics," *Free Inquiry* 22(1): Winter 2002.

McCormick, R. A. "Who or What is the Preembryo?" In Corrective Vision: Explorations in Moral Theology. Kansas City: Sheed & Ward, 1994.

McEwan PE, Gray GA, Sherry L, Webb DJ, Kenyon CJ. Differential effects of angiotensin II on cardiac cell proliferation and intramyocardial perivascular fibrosis in vivo. Circulation 1998;98:2765-2773.

McGee, Glenn, and Elizabeth Banger. "Ethical Issues in the Patenting and Control of Stem Cell Research." In Who Owns Life? Ed. David Magnus, Arthur Caplan, and Glenn McGee. Amherst, New York: Prometheus Books, 2002. 243-264.

McKenny, Gerald P. "Enhancements and the Ethical Significance of Vulnerability." In Enhancing Human Traits: Ethical and Social Implications. Washington, D.C.: Georgetown University Press, 1998. 222-37.

McKibben, Bill. Enough: Staying Human in an Engineered Age. New York: Henry Holt and Company, 2003.

McLean, Margaret. "Stem Cells: Justice at the Gate." Dialog, vol. 41, no. 3 (fall 2002).

Meilaender, Gilbert. "Terra es animata: On Having a Life." Hastings Center Report, vol. 23, no. 4 (July-August 1993): 25-32.

Murray, T., "Hard Cell," *The American Prospect,* September 24, 2001.

Ogawa M, Kizumoto M, Nishikawa S, Fujimoto T, Kodama H, Nishikawa SI. Blood 1999;93:1168-1177.

Okarma, T., Presentation at the September 4, 2003, meeting of the President's Council on Bioethics, Washington, D.C., available at www.bioethics.gov.

Okarma, T., Presentation at the September 4, 2003, meeting of the President's Council on Bioethics, Washington, D.C., available at www.bioethics.gov.

Onyango, P., *et al.*, "Monoallelic expression and methylation of imprinted genes in human and mouse embryonic germ cell lineages," Proceedings of the National Academy of Sciences of the *United States of America* 99: 10599-10604 (2002).

Orlic D, Kajstura J, Chimenti S, et al. Bone marrow cells regenerate infarcted myocardium. Nature 2001, 410:701-705.

Orr, R., "The Moral Status of the Embryonal Stem Cell: Inherent or Imputed?" *American Journal of Bioethics* 2(1): 57-59 (2002).

Pagano, S.F., *et al.*, "Isolation and characterization of neural stem cells from the adult human olfactory bulb," Stem Cells 18: 295-300 (2000).

Pevny, L., and Rao, M.S., "The stem-cell menagerie," Trends in Neurosciences 26: 351-359 (2003).

Pitt B, Segal R, Martinez FA, et al. Randomised trial of losartan versus captopril in patients over 65 with heart failure (Evaluation of Losartan in the Elderly Study, ELITE). Lancet 1997;349:747-752.

Pittenger, M.F., *et al.*, "Adult mesenchymal stem cells: Potential for muscle and tendon regeneration and use in gene therapy," Journal of Musculoskeletal and Neuronal Interactions *2: 309-320 (2002).*

Pursley, W. H., Presentation at the September 4, 2003, meeting of the President's Council on Bioethics, Washington, D.C., available at www.bioethics.gov.

Pursley, W.H., Presentation at the September 4, 2003, meeting of the President's Council on Bioethics, Washington, D.C., available at www.bioethics.gov.

R. Alta Charo, of the University of Wisconsin Law School, quoted in Coyle, M., "The Clone Zone," *The National Law Journal*, May 15, 2002.

Rafii S, Shapiro F, Rimarachin J, Nachman R, Ferris B, Weksler B, Moore AS, Asch AS. Isolation and characterization of human bone marrow microvascular endothelial cells: hematopoietic progenitor cell adhesion.

Rideout, W., *et al.*, "Correction of a genetic defect by nuclear transplantation and combined cell and gene therapy," Cell 109: 17-27 (2002).

Robertson, J., "Crossing the Ethical Chasm: Embryo Status and Moral Complicity," *American Journal of Bioethics* 2(1): 33 (2002).

Rohwedel, J., *et al.*, "Embryonic stem cells as an in vitro model for mutagenicity, cytotoxicity, and embryotoxicity studies: present state and future prospects," Toxicology In Vitro 15: 741-753 (2001).

Ryan, E.A., et al., "Clinical outcomes and insulin secretion after islet transplantation with the Edmonton protocol," Diabetes 50: 710-719 (2001).

Sato, N., *et al.*, "Molecular signature of human embryonic stem cells and its comparison with the mouse," Developmental Biology 260: 404-413 (2003).

Soria, B., *et al.*, "Insulin-secreting cells derived from embryonic stem cells normalize glycemia in streptozotocin-induced diabetic mice," Diabetes 49: 157-162 (2000)

Strong, C., "Those Divisive Stem Cells: Dealing with Our Most Contentious Issues," *American Journal of Bioethics* 2(1): 39-40 (2002).

Takahashi, T. et al. Ischemia- and cytokine-induced mobilization of bone marrowderived endothelial progenitor cells for neovascularization. Nat. Med. 1999; 5:434-438.

Tsai FY, Keller G, Kuo FC, Weiss M, Chen J, Rosenblatt M, Alt FA, Orkin SH. An early hematopoietic defect in mice lacking the transcription factor GATA- 2. Nature 1994; 371:221-225.

Weissman, I. L., "Stem Cells—Scientific, Medical and Political Issues," *New England Journal of Medicine* 346(20): 1576-1579 (2002).

Xu, R. H., *et al.*, "BMP4 initiates human embryonic cell differentiation to trophoblast," Nature Biotechnology 20: 1261-1264 (2002).

Zalzman, M., *et al.*, "Reversal of hyperglycemia in mice using human expandable insulin-producing cells differentiated from fetal liver cells," Proceedings of the National Academy of Sciences of the *United States of America 100: 7253-7258 (2003).*

Zhang M, Methot D, Poppa V, Fujio Y, Walsh K, Murry CE (2001)

Cardiomyocyte grafting for cardiac repair: graft cell death and anti-death strategies. J Mol Cell Cardiol 33:907-921.

Zhao, M., et al., "Amelioration of streptozotocin-induced diabetes in mice using human islet cells derived from long-term culture in vitro," Transplantation 73: 1454- 1460 (2002).

Zoloth, Laurie, "Jordon's Banks: A View from the First Years of Human Embryonic Stem Cell Research." In The Human Embryonic Stem Cell Debate. Ed. by Suzanne Holland, Karen Lebacqz, and Laurie Zoloth. Cambridge, MA: The MIT Press, 2001. 238.

Zulewski, H., *et al.*, "Multipotential Nestin-Positive Stem Cells Isolated From Adult Pancreatic Islets Differentiate Ex Vivo Into Pancreatic Endocrine, Exocrine and Hepatic Phenotypes," Diabetes 50: 521-533 (2001).

4

Biotechnology Research and Regulations

FINAL POLICY STATEMENT FOR RESEARCH AND REGULATION OF BIOTECHNOLOGY PROCESSES AND PRODUCTS

I. Introduction

The USDA portion of the "Proposal for a Coordinated Framework for Regulation of Biotechnology" (hereafter referred to as the December 3u, 1984 Notice) appeared at 49 FR 50897-50904. As a part of its policy perspective, USDA stated that agriculture and forestry products developed by biotechnology will not differ fundamentally from conventional products and that the existing regulatory framework is adequate to regulate biotechnology.

USDA has both research and regulatory responsibilities for biotechnology activities. This document provides significant new information in both areas. Section II describes 1985 Federal Register notices concerning USDA policies and responsibilities for biotechnology. Included in this discussion is an explanation of the assignment of responsibilities within USDA for the oversight of USDA funded research and for the regulation of the products of biotechnology. An understanding of the way in which USDA has divided these responsibilities should prove helpful to those in the private sector seeking review and/or approval of biotechnology applications.

A new section III has been added describing USDA's policy for agricultural biotechnology research. USDA is publishing as a companion document, USDA Guidelines for Biotechnology Research that will closely parallel the NIH Guidelines. The USDA guidelines will be issued under the authority of the Food Security Act of 1985 (Pub. L. 99-198). This Act amended section 1404(2) of the National Agriculture Research, Extension, and Teaching Policy Act (NARETPA). The Amendment gave the Secretary of Agriculture responsibility for establishing "appropriate controls with respect to the development and use of the application of biotechnology to agriculture." All USDA funded agriculture biotechnology research or research conducted at an entity receiving USDA funds would be subject to the USDA Guidelines for Biotechnology Research unless the specific research project is supported by and subject to the guidelines or regulations of another Federal agency. These Guidelines would encompass all phases of agricultural biotechnology research, i.e. (1) Contained laboratory experiments; (2) specialized isolation research (e.g., greenhouse, biotron); and (3) environmental research release (e.g., controlled and segregated field plots). USDA hopes that entities not required to comply with the Guidelines would voluntarily adhere to the requirements. To encourage compliance, USDA proposes to adopt the NIH policy of providing the researchers not required to comply with these Guidelines the opportunity to have their new biotechnology research proposals reviewed by USDA.

Those entities covered by the USDA Guidelines for Biotechnology Research would also be required to comply with any applicable statutes such as those set forth in section IV of this document, and any regulatory issues thereunder.

The Secretary of Agriculture has established an Office of Agriculture Biotechnology (OAB), which will have primary responsibility for implementing and coordinating the Department's policies and procedures pertaining to all facets of biotechnology. This includes the conduct of laboratory and field research, exprimentation on biotechnology products prior to their commercialization, and all matters of oversight of biotechnology in agriculture. The new office will report to the Assistant Secretary for Science and Education through the authority provided in the amendment to the Food Security Act of 1985. The Assistant Secretary for Science and Education will seek to establish an Agriculture Biotechnology Recombinant DNA Advisory Committee (ABRAC) and shall continue the responsibilities for agriculture formerly handled by the NIH-RAC during the last 10 years. The OAB shall operate in a close parallel manner to the Office of Recombinant DNA Activities (ORDA) of the National Institutes of Health. This includes the responsibility of the ABRAC and the implementation of the USDA Guidelines for Biotechnology Research. The NIH system is well respected both domestically and worldwide, and has achieved a high degree of efficiency in achieving broad confidence in the safety of new biological research conducted under its requirements.

The OAB also will serve as a focal point for coordinating a National Biological Impact Assessment Program, which is to evaluate and monitor the potential impacts of biotechnological processes and products on safety and the environment.

Section IV contains USDA's regulatory policy statements for veterinary biological products, plants and plant products, meat and poultry products, and seeds. USDA stated in the December 31, 1984 Notice that while its existing regulatory framework is adequate, it would constantly reevaluate its regulatory position and should additional regulatory measures become necessary, amend its regulations (49 FR 50904). For veterinary biologicals regulated under the Virus-Serum-Toxin Act (VSTA), USDA has identified three categories which may be derived by recombinant DNA techniques or developed from hybridomas. The categories are based on biological characteristics and safety concerns, and are described fully in section IV(A). The first category consists of inactivated recombinant DNA-derived vaccines, bacterins, bacterin-toxoids, virus subunits, or bacterial subunits, as well as monoclonal products. This category presents no new or unusual safety or environmental concerns. The second category includes those products containing live microorganisms that have been modified by the addition or deletion of one or more genes. Such products will be evaluated under current regulatory policies and procedures to assure that the addition or deletion of specific genetic information does not impart increased virulence, pathogenicity, or survival advantages. The third category included products using live vectors to carry recombinant derived foreign genes for immunizing antigens and/or other immune stimulants. Characteristics of safety and transmission must be established fully before questions and concerns dealing with safety to humans, animals, and release into the environment can be answered and before such products can be considered for licensing. Section IV(A) also includes new information about revised USDA review procedures for the importation of cell cultures and hybridomas. A brief discussion is included about the proposed regulations implementing the provisions of the amendments to the VSTA contained in the Food Security Act of 1985.

For organisms and products derived by the techniques of genetic engineering, USDA is proposing new rules to regulate organisms which are plant pests or which there is reason to believe are plant pests. It is USDA's policy to regulate certain genetically engineered organisms if the donor, vector/vector agent, or recipient organism is a member of a group of organisms that are known to contain plant pests, or if based on experience, USDA determines that a genetically engineered organism or product is a plant pest or if USDA has reason to believe that a genetically engineered organism or product is a plant pest. The proposed regulations are summarized in section IV(B).

The USDA policy for regulating meat and poultry products and seeds derived through biotechnology remains substantially as stated in the December 31, 1984 Notice, and appears in section IV (C) and (D).

A new section (V) has been added describing the scientific review mechanisms to be established by USDA to assit USDA Agencies in biotechnology research and regulatory decision-making. USDA has established a Committee on Biotechnology in Agriculture (CBA) chaired by the Assistant Secretary for Science and Education and the Assistant Secretary for Marketing and Inspection Services.

A detailed summary of comments on the December 31, 1984 Notice and USDA responses appears as section VI. The comments are organized to conform to the form of the December 31, 1984 Notice, with general comments and responses on the USDA regulatory philosophy followed by comments and responses on specific aspects of USDA's regulatory structure.

II. Notices

Three Federal Register notices concerning the

Department's biotechnology related activities have been published subsequent to publication of the December 31, 1984 Notice.

On July 19, 1985, a document amending the delegations of authority of USDA to assign responsibility for these Research and regulatory activities (7 CFR Part 2) was published in the Federal Register (50 FR 29367-29368).

In this document, the Secretary of Agriculture delegated responsibility to the Assistant Secretary for Marketing and Inspection Services to coordinate the development and carrying out of all matters and functions pertaining to the Department's regulation of biotechnology and to act as liaison on all matters and functions pertaining to the regulation to biotechnology between agencies within the Department and between the Department and government and private organizations. These responsibilities were further delegated from the Assistant Secretary for Marketing and Inspection Services to the Administrator of the Animal and Plant Health Inspection Service (APHIS).

Also in this document, the Secretary of Agriculture delegated responsibility to the Assistant Secretary for Science and Education to coordinate the development and carrying out of all matters and functions pertaining to agricultural research involving biotechnology conducted or funded by the Department including the development and implementation of guidelines for oversight of research activities, and to act as liaison on all matters and functions pertaining to agricultural research in biotechnology between agencies within the Department and between the Department and other governmental, educational and private organizations. n1

n1 The Assistant Secretary for Science and Education oversees the research activities of the Agricultural Research

Service (ARS), the Cooperative State Research Service (CSRS), the Extension Service (ES), and the Office of Grants and Program Systems (OGPS). The Assistant Secretary for Marketing and Inspecting Services oversees the regulatory activities of the Animal and Plant Health Inspection Service (APHIS), which includes Veterinary Services (VS) and Plant Protection and Quarantine (PPQ); the Agricultural Marketing Service (AMS); and the Food Safety and Inspection Service (FSIS). The policies and procedures of these agencies for biotechnology were described in the USDA portion of the coordinated policy statement at 49 FR 50899-50904

On September 23, 1985, USDA's APHIS published a notice which contained its policy statement and requirements for the control and protection of documents that contain confidential business information concerning biotechnology and the veterinary biologics program (50 FR 38561-38563).

On November 14, 1985, the Office of Science and Technology Policy published a notice in the Federal Register announcing the establishment of the Biotechnology Science Coordinating Committee (BSCC) (50 FR 47174-47195). This Committee is to serve as an interagency forum for coordinating science issues related to research and commercial applications of biotechnology. The notice also stated that USDA will establish a Committee on Biotechnology in Agriculture (CBA) to assit in assuring that research and regulatory decisions and based on the best science available.

III. USDA Research Policy Statement

USDA supports research to promote and protect the general health and welfare of the people of the United States. n2 Research program include: Studies on production of food and agricultural processing and marketing; identity and development of new crop and animal sources of food, fiber, and energy; increased agricultural efficiency and reduction of dependence on petroleum-based products; development of improved management and conservation of soil, water, forest, and range resources. The programs are fulfilled through State, Federal, and private industry cooperative efforts.

n2 See Addendum for Research Legislative Authorities.

In the areas of agricultural research relevant to biotechnology, many plant, animal, and microbial alterations have been developed for release through traditional genetic approaches such as mutagenesis and hybridization. In a complementary vein, beneficial introduction of organisms from abroad have established a sound base for research and regulatory oversight. The experience with these bases provide a substantial knowledge base for conducting evaluations of the safety and efficacy of biotechnology processes and products.

USDA will evaluate the environmental impacts in the context of individual experiments that encompass the entire range of experimentation from contained facilities to open field testing. As knowledge and experience are gained, broadly applicable procedures and guidelines will be developed. Particular consideration will be given to the stability of engineered changes and the possibility that genetic elements might be transferred from one organism to another. Also important will be the development of data that will enable predictions of which organisms may become established in new ecosystems, and resulting environmental consequences.

USDA considers products developed through biotechnological techniques as no different from those products resulting from research using conventional techniques providing appropriate research review is conducted with established protocols. Agricultural biotechnology research activities require appropriate review to avoid untoward effects on human health and the environment.

USDA expects to rely on the existing network of scientific expertise in the agriculture research community. Thousands of plant selections, animal breeding lines, and microorganisms are tested annually at sites under varying climatic conditions through the Nation. This network of scientific expertise permits continual, open assessment of agricultural research and products of that research in the field. USDA has broad statutory authority to conduct and support research in wide ranging areas of agriculture. In addition to the authorities described in the matrix of Federal Laws related to biotechnology found in the Federal Register Notice of November 14, 1985 (50 FR 47174-47195) the Food Security Act of 1985 (Section 1404(2) of the National Agriculture Research, Extension, and Teaching Policy Act Amendments of 1985, Pub. L. No. 99-198), made the Secretary of Agriculture responsible for establishing "appropriate controls with respect to the development and use of the application of biotechnology to agriculture." Through this authority, and pursuant to the Delegation of Authority Pertaining to Biotechnology published in the Federal Register on July 19, 1985 (50 FR 29367-68), the Assistant Secretary for Science and Education will complete development of a national system of agricultural biotechnology research oversight in much the same manner that agriculture has been a part for the last 10 years through the NIH-RAC.

The Assistant Secretary for Science and Education has initiated the establishment of the Agriculture Biotechnology and Recombinant DNA Advisory Committee (ABRAC), to be managed through an Office of Agriculture Biotechnology (OAB) which is a parallel to the National Institutes of Health Recombinant DNA Advisory Committee (NIT-RAC) and Office of Recombinant DNA Activities (ORDA). The

OAB will serve as the focal point for developing and coordinating USDA policies and activities pertaining to biotechnology research and will perform related interagency and public liaison functions. OAB will also assist in carrying out the responsibilities assigned to the Assistant Secretary for Science and Education, including the development and implementation of policies and procedures, and guidelines for the conduct of laboratory and field research.

All federally-funded agriculture biotechnology research or research conducted at an entity receiving USDA funds will be subject to the USDA Guidelines for Biotechnology Research, which are published as a companion document to this policy statement, unless the specific research project is supported by and subject to the guidelines or regulations of another Federal agency. These Guidelines encompass the entire spectrum of degrees of containment in agricultural biotechnology research i.e.: (1) Contained laboratory experiments; (2) specialized isolation research (e.g., greenhouse, biotron); and (3) environmental research agricultural biotechnology release (e.g., controlled and segregated field plots). Research investigators not required to comply with USDA Guidelines will be encouraged to follow these Guidelines. To assure consistency, USDA adopted the model established by the NIH of providing such researchers with the opportunity to have their biotechnology research proposals reviewed as required by the Guidelines.

The USDA Guidelines for Biotechnology Research require that research organization use the Institutional Biosafety Committee (IBC) concept as established by NIH. This requirement assures that each research organization and its investigators employ a multidisciplinary team to assit in carrying out their responsibilities under the Guidelines. The IBC's, as described in the Guidelines, would consist of persons with relevant agricultural expertise in areas such as recombinant DNA technology, biological safety, physical containment, and ecology. Requests for review beyond IBC should be sent to the Office of Agriculture Biotechnology (OAB) through the Assistant Secretary of Science and Education, Room 324-A, Administration Bldg., Washington, D.C. 20250.

These Guidelines also would require compliance with existing statutes of the USDA involving the movement of regulated organisms that require the issuance of a permit. The movement of microorganism injurious to plants and animals as well as the movement of certain non-indigenous plants and animals would continue to follow long-established procedures for USDA approval. After review, a permit, if needed, may be issued that allows movement. It is the responsibility of the research scientists to obtain that permit.

The Assistant Secretary for Science and Education will complete establishment of a National Biological Impact Assessment Program (NBIAP) as indicated in the USDA Guidelines for Biotechnology Research. NBIAP would serve to assist USDA in the evaluation and monitoring of biotechnology research and impact over time. Coordination of NBIAP will be provided through OAB.

IV. USDA Regulatory Policy Statements

The existing USDA regulatory authority for biotechnology was listed in the matrix of the December 31, 1984 Notice at 49 FR 50860-50874 and described in brief at 49 FR 50898-50899. The statutes considered most applicable to biotechnology applications are the Virus-Serum-Toxin Act (VSTA) of 1913 (21 U.S.C. 151-158), the Federal Plant Pest Act (FPPA) of May 23, 1957 (7 U.S.C. 150aa-150jj), the Plant Quarantine Act (PQA) of August 20, 1912 (7 U.S.C. 151-164, 166, 167), the Organic Act of September 21, 1944

(7 U.S.C. 147a), the Federal Noxious Weed Act (FNWA) of 1974 (7 U.S.C. 2801 *et seq.*), the Federal Seed Act (FSA) (7 U.S.C. 551 *et seq.*), the Plant Variety Protection Act (PVPA) (7 U.S.C. 2321 *et seq.*), the Federal Meat Inspection Act (FMIA) (21 U.S.C. 601 *et seq.*), and the Poultry Products Inspection Act (PPIA) (21 U.S.C. 451 *et seq.*).

A. Veterinary Biological Products

Under the Virus-Serum-Toxin Act of 1913, 21 U.S.C. 151-158, the USDA exercises regulatory authority over all veterinary biologics imported into the United States or shipped or delivered for shipment interstate. Recent amendments contained in the Food Security Act of 1985 have extended this authority to products which are shipped intrastate or exported, and have given the Department additional enforcement mechanisms such as the power to detain and seize products. Under the VSTA, veterinary biologics may not be shipped or delivered for shipment if they are worthless, contaminated, dangerous, or harmful. Veterinary biological products must be prepared in a USDA-licensed establishment under regulations promulgated by the Secretary of Agriculture. Those products which are imported into the United States must be imported under a permit issued by the Secretary. The pertinent regulations for veterinary biologics are found in Title 9 of the Code of Federal Regulations, Parts 101 through 117. New regulations will be drafted to implement the provisions of the amendments to the VSTA. Such regulations will provide for a more comprehensive regulatory scheme, including seizure and condemnation and detention procedures. They also will establish procedures to be used in the issuance of special licenses and exemptions provided for by the legislative amendments.

Veterinary biological products are defined in the

governing regulations, 9 CFR 101.2(w) as "all viruses, serums, toxins, and analogous products of natural or synthetic origin, such as diagnostics, antitoxins, vaccines, live microorganisms, killed microorganisms, and the antigenic or immunizing components of microorganisms intended for use in the diagnosis, treatment, or prevention of diseases of animals."

Licensing provisions for veterinary biological products and establishments are found in Part 102 of the USDA regulations (9 CFR Part 102). A product license requires the satisfactory completion of various requirements to assure purity, safety, potency, and efficacy of the products. The specific requirements were discussed in the December 31, 1984 Notice at 49 FR 50899.

Pursuant to § 103.3 (a) through (g) of the USDA regulations, a person may be authorized to ship unlicensed biological products for the purpose of evaluating experimental products by treating limited numbers of domestic animals if USDA determines that the conditions under which the experiment is to be conducted are adequate to prevent spread of disease and approves the procedures set forth in the request for such authorization (9 CFR 103.3 (a)-(g)).

Upon satisfactory completion of all requirements, including review and acceptance of labels, a U.S. Veterinary Biological Product License may be issued.

The application of new biotechnological procedures for the production of veterinary biological products is expanding constantly. For the purposes of licensing, biologics derived by recombinant DNA-techniques or developed from hybridomas, may be classified into three broad categories. This division is based upon the biological characteristics of the new products and the safety concerns they present, and is wholly analogous to the approach used in other veterinary biologics.

The first category includes inactivated recombinant DNA-derived vaccines, bacterins, bacterin-toxoids, virus subunits, or bacterial subunits. These nonviable or killed products pose no risk to the environment and present no new or unusual safety concerns. Monoclonal antibody (hybridoma) products used prophylactically, therapeutically, or as components of diagnostic kits also are included in this category.

The second category includes those products containing live microorganisms that have been modified by the addition or deletion of one or more genes. Deleted genes may code for virulence, oncogenicity, enzyme activity, or other biochemical functions. Added genes may result in the expression of new immunizing antigens or the production of novel biochemical byproducts such as beta-galactosidase. Precautions must be exercised to assure that this addition or deletion of specific genetic information does not impart increased virulence, pathogenicity, or survival advantages in these organisms which are greater than those found in natural or wild-type forms.

Modifications also must not impart undesirable new or increased adherence or invasion factors, colonization properties, or intrahost survival factors. It is important that genes added or deleted do not compromise the safety characteristics of the organisms. In most cases it is expected that they will be improved, and would therefore not pose any new threat to humans, other animal species, or to the environment.

The genetic information to be added or deleted must consist of well-characterized DNA segments. Required licensing data may include base pair analysis, sequence information, restriction endonuclease sites, as well as phenotypic characterization of the altered organism. A comparison is also required to be made between the genetically engineered organism and the wild-type form with respect to biochemical pathways, virulence traits, or other factors affecting pathogenicity.

The third category includes products using live vectors to carry recombinant-derived foreign genes that code for immunizing antigens and/or other immune stimulants. Live vectors may carry multiple recombinant-derived foreign genes since they can carry large quantities of new genetic information. They also are efficient at infecting and immunizing target animal species. These properties, for example, make vaccinia virus recombinants very popular subjects for vaccine development programs.

Live vectors currently being evaluated by licensees, applicants, and other research organizations include vaccinia, bovine papilloma virus, adenoviruses, Simian Virus-40, and yeasts. Characteristics of safety and transmission must be examined before questions and concerns dealing with safety to humans, animals, and release into the environment can be answered and before such products can be considered for licensing.

USDA will continue to avail itself of additional expertise from the Public Health Service "Interagency Group to Monitory Vaccine Development, Production, and Usage." This interagency committee will be utilized to consider potential human health hazards from the use of veterinary biological products and to review issues such as those arising from the potential effect of organisms potentially pathogenic to people or animals.

Veterinary biological products prepared using modern biotechnological procedures such as recombinant DNA, chemical synthesis, or hybridoma technology will be treated similarly to products prepared by conventional techniques.

The unlimited number and kind of products that may result from these modern biotechnology procedures make it impossible to define all requirements in specific terms. Each product is evaluated individually to determine what will be necessary to establish its purity, safety, potency, and efficacy. Scientific considerations may dictate generic areas of concerns or the use of certain tests for specific situations. Special assays, preferably using in vitro methods, may be required for potency and stability determinations. Additional tests may be required to assure safety, especially when live microorganisms are present in the biological products.

USDA is authorized to issue three types of permits for importing biological products into the United States (9 CFR 104.2). A separate United States Veterinary Biological Product permit is required for each shipment of biological product to be imported.

Permits are required for imported biological products used for research and evaluation, distribution and sale, or transit shipment only. Requests for application (U.S. Form 14-5) should be submitted to the Veterinary Biologics Staff, Veterinary Services, Animal and Plant Health Inspection Service, 6505 Belcrest Road, Hyattsville, Maryland 20782.

To provide guidance to current or prospective manufacturers employing modern biotechnological methods, the following points are presented:

1. *Recombinant DNA-Derived Products.* Genetic information coding for a product of interest and other sequences not indigenous to the host are referred to as foreign DNA. Recombinant DNA technology encompasses the isolation, characterization, and expression of foreign DNA in organisms or vectors. The specific cloned nucleotide segment coding for the desired product or other foreign

DNA segments must be defined in data supporting each license application. These data must also include a description of the source of the DNA and the nucleotide sequence.

A vector is a cloning vehicle which provides a suitable origin of replication necessary for production of foreign DNA. Such replicons may be derived from plasmids, bacteriophages or viruses such as vaccina, bovine pipillomavirus, adenoviruses, or SV-40.

Production of functional gene products depends on the efficient expression of cloned DNA-vector complexes in suitable host organisms. Tissue culture cells, bacteria, yeasts, and virus cells may be used as hosts for replication of vectors. The mechanisms of transfer, the copy number, and the physical state of the constructed vector inside the host cell, integrated or extrachromosomal, must be described.

USDA's licensing procedure for veterinary biological products derived from recombinant DNA involves a careful evaluation of each product on an individual basis to assure purity, safety, potency, and efficacy. Scientific and safety considerations may require specific safeguards and procedures in some situations. The USDA strongly recommends that all applicants establish Institutional Biosafety Committees which follow applicable provisions of the NIH Guidelines for Research Involving Recombinant DNA Molecules. USDA intends to propose guidelines which specifically relate to veterinary biological products. Amendments of the regulations and standards dealing with veterinary biologics will also be considered.

2. *Chemically Synthesized Antigens.* When the product consists of chemically synthesized polypetides, the appropriate amino acid sequences will mimic the antigenic site or epitope found in the native antigen where one exists. Procedures used to increase or prolong an immune response,

such as coupling to carrier proteins or addition of adjuvants, must also be described. Immunological data derived from chemically synthesized peptides must be as definitive as those from natural antigens.

3. *Monoclonal Antibody Products.* The specificity and potency of monoclonal antibody will be compared with those of similar policyclonal antibody products where appropriate. The sensitivity and specificity of monoclonal antibody products used in diagnostic test kits and their potency characteristics when used therapeutically must be similar to conventional antibody. Monoclonal antibody must be derived from Master Cell Stocks which meet the applicable requirements of 9 CFR 113.52. In addition, as is currently required, a description of cell cloning procedures, preparation, and characterization of cell passages must also be provided.

The Outline of Production must describe all processes including scale-up, ascites fluid or cell culture supernatant preparation, purification, concentration, and inactivation, Mouse colonies must be screened to demonstrate freedom from adventitious agents, especially those detected by the mouse antibody production (MAP) test. If the MAP test discloses the presence of adventitious agents, the product shall not be released unless inactivation procedures approved by Veterinary Services have been performed and tests conducted to ensure proper application of the procedures.

4. *Master Seeds.* Bacterial or viral seed stocks used to prepare veterinary biological products must meet established procedures used to certify Master Seeds for biological products.

The Master Seed for recombinant DNA-derived products may consist of a plasmid or virus carrying the inserted gene. This constructed plasmid is then introduced into the appropriate eukaryotic or prokaryotic expression system selected for vaccine production. Genomic DNA may also be transfected directly into a variety of mammalian cells. Alternatively, in such cases, the stable transfected cell could be considered as the Master Seed.

The establishment of Master Seeds consisting of constructed plasmids or transfected cells requires submission of background information concerning the recombinant DNA procedures used to isolate, purify, and identify genetic material from one source and the modification used for inserting of this material into a new host. Data from cloning, isolation, proliferation, and selection of genetically unique cells would be retained by licensed applicants. In order to characterize adequately the foreign DNA used to code for a particular antigen, the manufacturer must provide a nucleotide sequence analysis.

Tissue culture-propagated cells from vertebrate animals used for vector propagation and antigen production must meet the requirements of 9 CFR 113.51 or 113.52.

If a Master Seed has been accepted by Veterinary Services for use in a licensed product, further genetic modifications may be approved with reduced requirements for additional host animal efficacy studies.

Each Outline of Production must be prepared in accordance with 9 CFR 114.9. Outlines must include procedures to ensure consistency in production and recovery of specific antigenic material. Recovery procedures must include removal of excessive antibiotic levels (9 CFR 114.10) and undesirable fermentation byproducts such as excessive levels of bacterial endotoxins. Serial release tests for purity, safety, and potency will be required. In addition product characterization tests may be required to demonstrate consistent gene expression.

Organisms and Vectors

Pursuant to the Act of February 2, 1903, (21 U.S.C. 111), and the VSTA, USDA has authority to issue such regulations and take such measures as may be deemed proper to prevent the introduction or dissemination into the United States for the contagion of any contagious, infectious, or communicable disease of animals and/or live poultry from a foreign country into the United States or from one State or territory of the United States or the District of Columbia to another. The importation into the United States or interstate shipment of organisms and vectors is regulated under 9 CFR Part 122. Organisms and vectors are defined in 9 CFR 122.1 as entities which may introduce or disseminate any cantagious or infectious disease of animals. Such substances may not be shipped interstate or imported without a permit. Permit applications must completely describe the substances, intended use, location of the permittee, and safeguards.

A number of revised administrative and technical provisions have been instituted to expedite the USDA review and issuance of permits for importation or organisms and vectors which include cell cultures and hybridomas. No animal-origin biological materials, such as cell cultres, monoclonal antibodies, organisms, vectors, or related material, may be imported into the United States without a Veterinary Services (VS) Permit (VS Form 16-3A). To obtain a permit, an application (VS Form 16-3) should be submitted to: Import-Export Staff, Organisms and Vectors, VS, APHIS, USDA, 6505 Belcrest Road, Hyattsville, MD 20782. This is different from the permit required to import veterinary biologics pursuant to Part 104 of the USDA regulations governing such products (VS Form 14-5 and 14-6).

Applicants must also complete the questionnaire entitled "Importation Information" and submit it with their application. Based upon the information submitted by the applicant, a determination will be made if the material to be imported requires safety testing to ensure it is free from livestock pathogens. Safety testing is conducted at the Foreign Animal Disease Diagnostic Laboratory (FADDL), Plum Island, New York.

Applicants will be advised if a safety test is required and will be given an estimate of the cost for conducting the test. Applicants desiring to have material safety tested must enter into a Cooperative Trust Fund Agreement with APHIS, VS, and deposit in advance sufficient funds to cover the estimated cost. The Import-Export Animals and Products Staff will initiate the Cooperative Trust Fund Agreement. In order to expedite the procedure, VS may issue a permit for the material to be shipped to FADDL pending receipt of the funds and Cooperative Trust Fund Agreement. However, the signed Cooperative Agreement, plus the necessary funds, must be received by VS before testing can be scheduled at FADDL.

Usually 60 to 90 days is needed for issuing a permit for importing material to Plum Island, New York, the completion of safety tests, and the transfer of the imported material to the applicant. A minimum of four vials, each containing at least 1 million cells from a uniform lot, is required for the safety testing.

When the test is completed and a determination made that the imported material is free from livestock pathogens, the remainder of the imported material is released directly to the importer under conditions specified in the permit.

If an importer wishes to import cell cultures and/or

hybridoma cells on a regular basis, the applicant may enter into a continuous Cooperative Trust Fund Agreement with VS and establish an escrow account to ensure that unnecessary delays will not occur due to insufficient funds.

Each safety test utilizing susceptible host animals usually cost approximately $2,000 to $3,000. Sometimes it is possible to reduce the cost by pooling samples in one host animal test. Scientists at FADDL developed in vitro safety tests to detect certain livestock pathogens resulting in substantial cost savings for importers. The current cost of each in vitro test is approximately $500, depending upon the type of animal disease present in the country of origin as well as the intended use of the imported material.

Safety testing may not be required for some cell cultures imported for human diagnostic purposes and research. Examples of material which could enter without safety testing include cultured human bone marrow cells, amniocentesis samples, and cells imported for karyotype analysis. Applications for such cell cultures will be considered individually.

Permit applications are evaluated by a new classification scheme that correlates intended use of imported cell cultures with the level of safety testing conducted at FADDL.

The following classification of cell cultures is based on intended use and generally indicates the level of safety testing required.

Class I Cell cultures to be used for the production of products such as vaccines, hormones, or other biologicals to be used in livestock, poultry, or for commercial distribution.

Requirement: These cell cultures must be safety tested at FADDL using susceptible host animals, approved in vitro test, and/or laboratory animals.

Class II Cell cultures to be used only for in vitro studies and not to be used in animals other than primates.

Requirement: These cultures may not require safety testing. The material may be sent directly to the importer when no safety testing is required. The permit (VS Form 16-3A) will specify restrictions such as "FOR IN VITRO LABORATORY TESTS: DO NOT INOCULATE INTO LIVESTOCK, BIRDS, OR LABORATORY ANIMALS."

Cell cultures imported under permit which do not require a safety test may not be distributed to other laboratories without prior approval from USDA, APHIS, VS. Applications for the distribution of imported material should be submitted to the USDA, APHIS, VS, Import-Export Staff, Organisms and Vectors.

When appropriate, a review is conducted by the Administrator's Parent Committee on Organisms and Vectors. Members of this committee have wide expertise in evaluating safety. Clearance may also require testing in high security facilities at the Veterinary Services, FADDL, Plum Island, New York.

B. Plants and Plant Products

Pursuant to the authority granted by the Federal Plant Pest Act (FPPA) of May 23, 1957, as amended (7 U.S.C. 150 aa through 150 jj), and the Plant Quarantine Act (PQA) of August 20, 1912, as amended (7 U.S.C. 151 through 164, 166, and 167). USDA has regulatory authority over the movement into or within and through the United States of plants, plant products, plant pests, and any product or article which may contain a plant pest at the time of movement. These articles are regulated in order to prevent the introduction, spread, or establishment of plant pests new to or not widely prevalent in the United States. The regulations implementing this statutory authority are found in 7 CFR Parts 300 through 399.

"Plant Pest," as defined by statute, means any living stage of any insects, mites, nematodes, slugs, snails, protozoa, or other intertebrate animals, bacteria, fungi, or parasitic plants or reproductive parts thereof, viruses, or any organisms similar to or allied with any of the foregoing, or any infectious substances, which can directly or indirectly injure or cause disease or damage in any plants or parts thereof, or any processed, manufactured, or other products of plants (7 U.S.C. 150aa(c)).

"Movement," as defined by statute, means to ship, deposit for transmission in the mail, otherwise offer for shipment, offer for entry, import, receive for transportation, carry, or otherwise transport or move, or allow to be moved, by mail or otherwise (7 U.S.C. 150aa(g)).

The current permit system requirements for the movement into or within and through the United States of plants, plant products, plant pests, and other articles regulated by FPPA and PQA were fully described in the December 31, 1984 Notice at 49 FR 50900-01. The procedures for issuing permits for the movement of plant pests were discussed separately from plants, plant products and other articles which may contain plant pests at 49 FR 50901-02. USDA regulates the importation of noxious weeds through a permit system similar to that established for plant pests. The existing regulations in 7 CFR Part 360 which designate plants as noxious weeds and establish procedures for obtaining an import permit were described at 49 FR 50902.

Regulation of the Introduction of Organisms and Products Altered or Produced Through Genetic Engineering Which Are or Which There is Reason to Believe Are Plant Pests

The FPPA and PQA are applicable to the movement of plants, plant products, and other articles and plant pests developed through genetic engineering if such plants, plant products, other articles, or plant pests present a risk of plant pest introduction, spread, or establishment.

Under the authority granted by the FPPA and PQA, USDA is proposing new regulations which would impose restrictions on the introduction of organisms and products altered or produced through genetic engineering which are plant pests or which there is reason to believe are plant pests.

In accordance with the provisions of the FPPA and PQA. USDA must determine the plant pest status of plants, plant products or articles to be moved into or within or through the United States. The evaluation process for determining what safeguards, if any, can be imposed which would allow the movement of the plant pest without risk that the plant pest would be disseminated were described in the December 31, 1984 Notice at 49 FR 50901-02. For genetically engineered material from dissimilar source organisms (inter-generic combinations), the determination may be complex. Information about genetically engineered organisms produced through the use of donor, vector/vector agent and recipient organisms that are from a list of known plant pests is needed in order that such organisms be properly regulated.

During the past year, USDA has received permit applications to move genetically engineered organisms into or through the United States. USDA is confident that organisms altered through genetic engineering will play a major role in increased plant yield and improved plant quality. However, a genetically engineered organism derived from organisms that are plant pests also presents a risk of plant pest introduction. The organisms themselves, the cultures in which they are transported, or their packaging may be contaminated with plant pathogens. Genetic alteration may create a plant pest new to and not widespread in the United States. It is necessary, therefore, to establish appropriate safeguards to prevent the introduction of genetically engineered organisms that pose a threat to agriculture. Other genetically engineered organisms that are not plant pests or where there is no reason to believe such organisms are plant pests would not be regulated.

New data have to be required in order to properly evaluate permit applications for those organisms which are plant pests or which there is reason to believe are plant pests. A determination was made that additional data requirements would be incorporated into proposed regulations for those genetically engineered organisms which are of concern under the provisions of the FPPA and PQA.

USDA is publishing as a companion document in the "proposed rules section" of this issue of the Federal Register its proposed regulations pertaining to organisms and products altered or produced through genetic engineering which are on plant pests or which there is reason to believe are plant pests.

The proposed regulations would establish a new part entitled, "Introduction of Organisms and Products Altered or Produced Through Genetic Engineering Which Are Plant Pests or Which There is Reason to Believe are Plant Pests", in Title 7 of the Code of Regulations (7 CFR), pursuant to the authority of the FFPA, as amended (7 U.S.C. 150aa-150jj) and the PQA, as amended, (7 U.S.C. 151-164, 166, 167). Such proposed regulations would regulate the importation into and movement within and through the

United States as well as prevent the release into the environment of certain organisms, or products altered or produced through genetic engineering, which are plant pests or which there is reason to believe are plant pests.

The proposed regulations would restrict the "introduction" of certain organisms and products altered or produced through genetic engineering, referred to as "regulate articles." In this context, "introduction" means to move into the United States, to release into the environment, or to move interstate, or any attempt thereat." "Release into the environment" means "use of a regulated article outside the constraints of physical confinement that are found in a laboratory, contained greenhouse, or fermenter or other contained structure."

USDA's proposed regulations, which are designed to prevent the release into the environment of genetically engineered organisms which are plant pests or which there is reason to believe are plant pests are consistent with the legislative intent of the FPPA. The FPPA was enacted in 1957 and was intended as "gap filing" legislation for the purpose of protecting American agriculture against invasion by plant pests and diseases which are new to or not theretofore known to be widely prevalent or distributed within and throughout the United States. The FPPA also provides USDA with authority to regulate insects or pests that might later be found to be injurious to cultivated crops. The release into the environment of a genetically engineered plant pest is tantamount to the introduction of a plant pest which is new to and not theretofore known to be widely prevalent within and throughout the United States and subject to regulation under the FPPA.

It should be noted that "regulated article" would be defined as any organism or product altered or produced through genetic engineering, if the donor organism, recipient organism, or vector or vector agent belongs to a group of organisms designated by the proposed regulations as having plant pests or any organism or product which USDA determines is a plant pest or which there is reason to believe is a plant pest. Under USDA's proposed definition, certain microorganisms would be excluded if the recipient microorganism is non-pathogenic, is non-infectious, and otherwise not a plant pest, and resulted from the addition of genetic material that is well characterized and contains only non-coding regulatory regions. Restrictions would be required for regulated articles because they are plant pests, or because USDA has reason to believe they are plant pests. The proposed regulations would require that a person obtain a permit prior to the introduction of a regulated article and would list specific conditions required for the introduction of a regulated article. The regulated article could be introduced only if all conditions in the proposed regulations as well as all conditions specified on the permit were met. It is important to note that in considering whether a permit can be issued for the introduction of a genetically engineered organism, USDA will perform the same comprehensive analysis that is used in determining whether a permit can be issued for the movement of a "conventional" plant pest. Such asessment shall include an examination of the factors that were discussed in the December 31, 1984, Notice at 49 FR 50901-02 as part of the evaluation process for determining what safeguards can be imposed which would allow the movement of a plant pest without risk of dissemination. These factors are oriented toward an examination of the ecological and environmental effects of a release of the genetically engineered organism or product into the environment.

The proposed regulations also contain provisions for a certificate of exemption for those organisms or products altered or produced through genetic engineering that are not subject to the proposed regulations. A person seeking to introduce an exempt article could voluntarily request a certificate of exemption to facilitate the introduction of the organism or product.

The proposed regulations provide a list of groups of organisms which are plant pests or contain plant pests. If the donor, vector/vector agent, or recipient of the genetically engineered organism is derived from an organism on the list of organisms containing plant pests, such genetically engineered organism would be deemed a "regulated article".

As defined in the proposed regulations, a plant pest includes microorganisms such as bacteria and viruses, and thus a "regulated article" may be a microorganism unless it meets the provisions for exclusion. It is important to note that in some instances certain microorganisms will be subject to joint regulation by USDA and EPA. USDA has jurisdication over certain microorganisms under the FPPA and PQA if the microorganisms are a plant pest. EPA would have jurisdiction under the Toxic Substances Control Act (TSCA) if the microorganism is deemed to be a "new" microorganism or under the Federal Insecticide, Fungicide, and Rodenticide Act, as amended (FIFRA) if the microorginism is to be used as a pesticide. Because each Agency has a different statutory mandate, certain jurisdictional overlaps cannot be avoided. However, EPA and USDA will work cooperatively and simultaneously in the evaluation of genetically engineered microorganisms that fall under the jurisdication of both Agencies. To expedite the review of these microorganisms each Agency will appoint contact persons to coordinate the review to ensure data requests are not duplicated.

The specifics of which microorganisms will be subject

to dual Agency review, or primarily single Agency review, is set forth in the preamble of USDA's proposed regulations being published as a companion document to this policy statement. That document should be consulted for further information.

A key to determining whether a genetically engineered organism will be regulated by USDA is the list of organisms containing plant pests that appears in § 340.2 in proposed Part 340. USDA acknowledges that this is not an exhaustive list, and that it does not attempt to list every pest species. Comments are welcome on the list as well as on other parts of the proposed regulations.

In order to solicit as many comments as possible on the list and all other parts of the proposed regulations, USDA has scheduled public hearings in Washington, DC and Sacramento, California, during the 60-day comment period. The time and place of the public hearings as well as the address to send written comments is specified in the preamble to the proposed regulations.

USDA believes that through the submission of detailed comments and full participation by public and private interests, USDA will be able to promulgate a final regulation that will prevent the introduction and dissemination of genetically engineered organisms which are plant pests or which there is reason to believe are plant pests, yet not impede the development of biotechnology.

C. Meat and Poultry Products

The Food and Safety Inspection Service (FSIS) is responsible for assuring the safety, wholesomeness, and proper labeling of food products prepared from domestic livestock and poultry. The Federal Meat Inspection Act (FMIA) and the Poultry Products Inspection Act (PPIA) require FSIS to inspect cattle, sheep, swine, goats, equines, poultry, and food products prepared from them which are intended for use as human food to assure that they are wholesome, not adulterated, and properly labeled, marked, and packaged. Inspection under these statutes is mandatory. The cost of inspection, except for overtime and holiday inspection work, is required to be borne by the USDA. Food, animals and animal products, other than those required to be inspected under the FMIA and PPIA, may be inspected under a voluntary, reimbursable inspection program established under the Agricultural Marketing Act of 1946.

Within the framework of food safety statutes, FSIS has developed regulations for research on animals that are administered experimental animal drugs, biologics, and pesticides (9 CFR 309.17 and 381.75). These regulations state that no animal used in any research investigation involving an experimental biological product, drug, or chemical shall be eligible for slaughter at an official establishment unless certain conditions are met. These conditions include any of several different ways of demonstrating that the use of such biological product, drug, or chemical will not result in the products of such animals being adulterated.

Products Subject to Review. FSIS anticipates that many food animals which are subject to the new techniques of modern biotechnology will not differ substantially in appearance, behavior, or general health from currently inspected cattle, sheep, swine, goats, equines, and poultry. They would be subject to the same inspection procedures and regulations as tradionally inspected food animals. FSIS is aware that some genetically engineered animals, such as mosaics, chimeras, and some hybrids, may differ substantially from animals that are inspected currently under the FMIA and PPIA. If such animals are ever intended for use as human food and are presented for inspection at an official establishment, a decision would have to be made as to whether such animals were covered under the FMIA or PPIA, and if not, whether the FMIA and PPIA should be amended to require inspection of such animals and their products.

Implementation of Review Authority. FSIS's approach toward the review of food animals resulting from the techniques of modern biotechnology consists, in general, of two phases. The first, an experimental phase, focuses on the experimental aspects of vector administrative, gene transfer and gene expression. Since artificial vectors used in animal gene transfer may be considered as either animals drugs or animal biologics, their administration to food animals would be covered under the current regulations on animals used for research (9 CFR 309.17 and 381.75). The requirement that an animal carcass intended for use as human food not be adulterated may require that certain phenotypic, biochemical, and microbiological parameters not be exceeded before the animal can be slaughered for human food. Depending on future developments, FSIS may amend the regulations (9 CFR 309.17 and 381.75) to provide further assurance that the products of animals genetically engineered by certain techniques are not adulterated. The second phase would be carried out under existing regulations (9 CFR Parts 301 through 381) and would focus on the commercial development, production, inspection and labeling of food animals and food animal products.

D. Seeds

The Federal Seed Act (FSA) (7 U.S.C. 1551 *et seq.*) defines USDA regulatory authority over the importation and interstate shipment of agricultural and vegetable seeds.

It does not apply to the production or intrastate distribution of seeds or to seeds other than agricultural or vegetable seeds ("agricultrual seeds" are grass, forage, and field crop seeds).

The FSA prohibits interstate shipment of seed that contains noxious weed seeds at levels in violation of the laws of the State of destination or in excess of levels allowed by the Secretary of Agriculture. This provision applies primarily to seed adulterated with noxious weed seed. In a few instances, however, States have determined that a particular variety of agricultural or vegetable seed is itself a noxious weed. In these instances, FSA prohibits the interstate shipment of the seed into those States. The FSA also allows the Secretary to prohibit the importation of agricultural and vegetable seed which is adulterated with noxious weed seed or which is unfit for seeding purposes.

The authority granted to the Secretary by the FSA to prohibit the interstate shipment or importation of seeds which are found to be detrimental to the agricutural interests of the United States applies to seeds genetically engineered with the modern biotechnology to the same extent as any other seeds.

V. Scientific Review Mechanisms

The manner in which both regulation and oversight of research in agriculture-related biotechnology evolves and is implemented in the United States will have a direct impact on the competitiveness of U.S. industry in both domestic and world markets. Inconsistent or unnecessary procedures for regulation and research will place the U.S. scientific effort and U.S. producers at a substantial disadvantage. It also is important that safeguards be built into biotechnological research processes, and that releases be based on careful evaluations while further experience is being gained. Therefore, USDA feels that such regulatory and research decisions must be based on the best science available.

While the responsibilities within USDA for biotechnology reside with the Assistant Secretary for Science and Education and the Assistant Secretary for Marketing and Inspection Services as the delegates of the Secretary of Agriculture, in carrying out their respective responsibilities based on the best science available, they would be able to take advantage of the expertise and perspectives within the Federal Government through a committee to be called the Committee on Biotechnology in Agriculture (CBA). The CBA, to be chaired by these two Assistant Secretaries, will function both as a policy body in the USDA and a bridge between its research and regulating structures.

Committee on Biotechnology in Agriculture

The objectives of the CBA will include:

To provide advice, when requested, on initiatives, proposals, and policy for agriculture-related regulation and research, and assist in the coordination of these activities;

To review scientific issues submitted by agencies within the Department;

To assist in identifying data gaps for basic research in agricultural biotechnology;

To foster public awareness of the scientific issues in biotechnology;

To provide Departmental support for participation in the FCCSET BSCC.

USDA expects that the CBA also will utilize existing cooperative entities (e.g., other Federal agencies, universities,

State regulatory officials, the public sector, and industry) to acquire, when necessary, information for addressing those issues submitted to it. Such entities, when requested, can provide technical support for sound regulatory and research decisions regarding the use of biotechnology in agriculture and foresty. These entities offer a vast scientific resource upon which USDA can draw.

VI. Summary of Comments

USDA received the comments of one hundred-two (102) respondents, one-half of whom commented specifically on the USDA policy statement. Although USDA agencies considered all comments on the coordinated policy proposal, this response is confined to comments on the USDA portion of the notice.

The two largest categories of respondents were business and academic, followed closely by associations representing these interests. Comments came in lesser numbers from environmental and public interest groups, individuals, law firms, and foreign governments, as well as the National Institutes of Health Recombinant DNA Advisory Committee (NIH-RAC) and a member of the U.S. Congress

The USDA response to the comments follows the form of the original notice, with a discussion of comments on regulatory philosophy followed by a response to comments on the regulatory framework.

Comments on the Nature of Products of Modern Biotechnology: Fourteen respondents stressed their agreement with the USDA statement that "agriculture and forestry products developed by modern biotechnology will not differ fundamentally from conventional products," while six commenters dissented. Three respondents felt that genetic engineering across species barriers did create a potentially different product and the possiblity of unique ecological effects. Concern about the "need for public trust" and public assurance on safety and ethical issues was stressed by three commenters. Seven respondents agreed with USDA that "to date, no unique or safety problems have been associated with products of genetic engineering," but four of the same commenters who view biotechnology products as fundamentally different from conventional products stressed that the potential exists for safety problems with biotechnology applications.

Response: USDA recognizes the importance of ecological effects and the need for developing procedures responsive to public concerns about safety.

Although USDA's regulatory philosophy remains as stated, additions to regulatory procedures are being proposed for genetically engineered plants and plant products and veterinary biologics produced by biotechnology (see section IV). The previously discussed delegations of authority within USDA for biotechnology increase the effectiveness of the administration of current and proposed regulatory procedures affecting the products of modern biotechnology. For veterinary biological products, USDA is currently developing additional procedures pursuant to the VSTA, as amended, for evaluating requests to conduct experimental field trials with live vectors containing genetically engineered organisms or to support product license applications. The procedures being developed consider the parental organism and the effect of the gene alteration on the genetic properties of the recipient, especially the survival, reproduction, and dispersal characteristics. A careful analysis of the genetics, biology, and ecology of the wild-type and modified microorganisms will provide as reasonable prediction of the risks which might be associated with use of the altered organisms.

USDA is proposing regulations pursuant to the Federal Plant Pest Act (FPPA) and the Plant Quarantine Act (PQA) for regulating the introduction of certain organisms of products thereof altered or produced through biotechnology which are plant pests or may become plant pests. This proposed rule should assist USDA in assessing the ecological effects of the release of such genetically engineered organisms into the environment.

Guidelines for oversight of agricultural biotechnology research funded by USDA will be issued under the authority of the Food Security Act of 1985.

USDA also is establishing scientific review mechanisms to assist in research and regulatory decisions (see section V).

These proposed modifications in the procedural framework are described as a part of the final policy statement for veterinary biologics, plants and plant products, research, and scientific review mechanisms.

Comments of the Adequacy of Existing Authority: Thirteen commenters agreed with USDA that its existing regulatory framework is adequate for biotechnology applications, and nine favor the case-by-case approach under existing authority. Five commenters felt that new legislation is or may be needed; two of the five oppose the case-by-case approach.

Response: USDA has examined its statutory authority for regulating biotechnology products and processes, and USDA agencies have processed licensing and permit applications under the existing statutes. The existing authority is considered adequate at this time. Established procedures, with the proposed modifications, can be adapted effectively to handle biotechnology applications. USDA is currently considering genetic engineering applications on a case-by-case basis using existing authority.

Comments on Need for Procedures and Guidelines: Sixteen respondents commented that USDA had not outlined procedures for the review and approval of genetically engineered products. Twelve respondents stressed the need for flexibility, and six requested sunset provisions in USDA biotechnology regulations.

Response: The USDA policy statement of December 31, 1984, did outline procedures currently used for the review and approval of certain genetically engineered products. In considering license applications for genetically engineered veterinary biologics, USDA follows the standards and procedures applicable to all such products found in § § 101-117 of the applicable regulations and standards (9 CFR 101-117). In the December 31, 1984 Notice, USDA offered supplementary guidelines for licensing such products. New procedures are being developed to evaluate production and testing of veterinary biologics derived through use of genetic engineering techniques. The information needed for proper evaluation will depend on the parent organism and the effect of the gene alteration on the genetic properties of the recipient. A paper describing the USDA licensing policy for biologics produced by recombinant DNA technology was presented at the Joint International Association of Biological Standardization/World Health Organization Symposium on "Standardization and Control of Biologics Produced by Recombinant DNA Technology," Geneva, Switzerland, 1983 (published in *Developments in Biological Standardization,* V. 59, pp. 167-173, S. Korgel, Basel, 1985). The paper describes requirements for plasmid/vector characterization and stability, and correlation to conventional Master Seed concepts, as well as methodology which can be used to monitor antigenic expression, concentration, purification, and stability testing during production and recovery.

The movement of genetically engineered products which are plant pests and present a risk of plant pest introduction or spread is regulated by 7 CFR 330.200 implemented pursuant to the FPPA and PQA. The movement of organisms and vectors which may cause desease in animals is regulated under 9 CFR Part 122.

USDA realized that the statement left unanswered some questions about the means for review and approval of various genetically engineered products. The proposed regulations described in section IV(B), implemented under the authority of the FPPA establish permit requirements for the "introduction" of organisms altered or produced by genetic engineering which are or may become plant pests. The regulations would be flexible because organisms determined not to be plant pests would be exempt, and this category could be expanded in the future to include organisms whose plant pest status is currently uncertain and therefore restricted. It is hoped that the discussion in section IV(B) of this policy statement answers any remaining questions about the review and approval procedures for such genetically engineered products.

Comments on Confidential Business Information (CBI): Six commenters representing business and scientific interests expressed concern about the protection of "confidential business information" in the USDA regulatory process while two public interest groups stressed the "public's right to know."

Response: The USDA regulations implementing the Freedom of Information Act (FOIA) (5 U.S.C. 552) are found in 7 CFR 1.1-1.16. The FOIA provides that Federal agencies must make available to the public all records not specifically exempt from disclosure. Exemptions include "trade secrets and commercial or financial information." (5 U.S.C. 552(b)(4)). On September 23, 1985, USDA's APHIS issued a policy statement on the protection of privileged or confidential information (50 FR 38561-38563). This policy statement establishes requirements for the control and protection of documents received by APHIS that contain privileged or confidential business information concerning biotechnology and the veterinary biologics program. The procedures established conform to the FOIA requirements for both protection and disclosure.

Comments on Use of NIH Guidelines: Four respondents questioned the USDA requirements that manufacturers of veterinary biological products using recombinant DNA technology follow the National Institutes for Health Guidelines for Research Involving Recombinant DNA Molecules (NIH Guidelines). One respondent thought USDA implied that all people "who work with recombinant DNA plants" would be required to comply with the NIH Guidelines, and requested procedural guidelines for industry.

Response: The USDA does not require that the manufacturers of veterinary biological products or plant products of recombinant DNA technology follow the NIH Guidelines. However, USDA strongly recommends that all license applicants for veterinary biologics follow appropriate provisions of the NIH Guidelines, such as those regarding the establishment of an institutional biosafety committee. USDA intends to propose guidelines that will parallel closely the NIH Guidelines, and it intends to recommend strongly that entities not required to follow the USDA guidelines do so voluntarily.

Comments on Importation of Cell-Lines: Three

associations representing biotechnology companies requested that USDA take steps to reduce delays in the clearance and testing procedures required for the importation of biotechnology-derived products and cell-lines. On February 12, 1985, the Association of Biotechnology Companies (ABC) delivered a report on USDA importation guarantine issues to the APHIS Parent Committee for Foreign Pathogens and Vectors. This report was an attachment to the ABC comment letter.

Response: The USDA has instituted a number of revised administrative and technical provisions to expedite the issuance of permits for importation of organisms and vectors which include cell cultures and hybridomas. A supplementary questionnaire, designed to insure adequate information on cell cultures and products from recombinant DNA or hybridoma technologies, now accompanies each permit application. Applicants are advised whether or not a safety test is required and a cost estimate is given. Safety testing may be conducted concurrently with the administrative review of the permit application, but only at APHIS' Foreign Animal Disease Diagnostic Laboratory (FADDL) at Plum Island, New York. New test procedures have reduced the cost of safety testing, and the cost per sample can be further reduced by conducting a safety test with pooled samples. Permit applications are evaluated using a new classification scheme that equates intended use of imported cell cultures with the level of safety testing required at FADDL. Class I cell cultures, employed in the preparation of products such as enzymes, vaccines, or hormones for commercial use, are subject to complete safety testing. Class II cell cultures, used only for in vitro studies and not to be used in animals other than primates, are subject to a lesser degree of testing.

Comments on Risk Analysis: Seven respondents discussed

the issue of risk assessment or risk/benefit analysis of biotechnology applications. Comments varied from a recommendation that "standard risk assessment methodologies" be adopted by all agencies to a warning against attempting to regulate the "hypothetical and imaginary "potential' dangers" of recombinant DNA techniques.

Response: The National Environmental Policy Act (NEPA) applies to USDA actions. The "APHIS Guidelines Concerning Implementation of NEPA Procedures" (44 FR 50381, August 28, 1979) would be used to make an environmental assessment or environmental impact statement of the effects of a proposed release of a genetically engineered organism regulated pursuant to the VSTA, the FPPA and PQA, and related statutes. A formal risk management procedure based on a wide variety of safety concepts will be used to evaluate systematically proposed releases. The information required by any new regulations promulgated under the FPPA and PQA would be used to prepare the environmental assessment for release of a genetically engineered product which is a plant pest or may become a plant pest.

In normal husbandry and laboratory practices, veterinary biological products normally are not considered to be released into the environment. In the event that a conventionally prepared or recombinant derived product would be considered to be released into the environment, the issuance of a license or import permit would require compliance with procedures being developed and interagency approval. The procedures under development consider the parental organism and the effect on the gene alteration on the genetic properties of the recipient, especially the survival, reproduction and dispersal characteristics.

Safety, ethics, and policy issues in agricultural biotechnology research will be overseen by the Committee on Biotechnology in Agriculture (CBA) and such supporting technical advisory groups as may be established by the USDA agencies. Currently, all USDA and USDA-sponsored research involving recombinant DNA must be cleared prior to initiation for compliance with the NIH Guidelines.

Comments on Jurisdiction: The potential for overlapping jurisdiction in the policy notice drew the largest number of comments. Eighteen respondents pointed out that both USDA and EPA propose to regulate agricultural microorganisms. Respondents representing the interests of the veterinary biologics industry contended that a jurisdictional dispute between USDA and FDA delayed the approval of bovine interferon. While generally supporting the concept of the memorandum of understanding (MOU) between USDA and FDA to resolve jurisdictional disputes, one respondent challenged the legality of the MOU, noting that it contains the statement that "animal biological products generally act through a specific immune process," while USDA's current regulations do not restrict its jurisdiction to products operating through such a mechanism of action. Industry respondents also pointed out that the intrastate producer of veterinary biologics is not regulated by USDA. Two firms and one industry association urged prompt Federal oversight action so that States do not act independently to regulate biotechnology products.

Response: USDA agrees that there is the potential for overlapping jurisdiction among the Federal agencies involved in regulating biotechnology products. USDA and EPA representatives have discussed jurisdiction over genetic engineering applications since 1983. USDA and EPA have begun to establish a regulatory procedure for reviewing certain submissions of genetically engineered microorganism applications, a procedure which has resulted in joint consultation on several proposals for release into the environment of organisms altered by genetic engineering.

For veterinary biologics regulated under the VSTA, use of procedures currently under development will increase USDA effectiveness in evaluating biotechnology license and product applications. The MOU between USDA and FDA was published on June 8, 1982, in an attempt to resolve the issue of new products which fall into the questionable definitional area between animal drugs regulated by FDA and animal biologics regulated by USDA. An interpretation by some that the term animal biologics only includes substances that act through a specific immune process has resulted in some confusion. There is nothing in USDA's current regulations or law which restricts its jurisdiction to products acting solely through this mechanism of action, and because of this fact, the memorandum qualifies its reference to specific immune process by the word "generally." Although efforts will be made to clarify the issue further, it should be noted that there appears to be little uncertainty about whether a particular product is a veterinary drug or biologic.

The Food Security Act of 1985 contains amendments to the VSTA that extend USDA's jurisdiction to veterinary biologics which are shipped intrastate or exported. The provisions of the amendments are discussed more fully in Section IV.

Comments on the National Biological Impact Assessment Program (NBIAP): Seven respondents commented on the NBIAP, the proposal by the National Association of State Universities and Land Grant Colleges (NASULGC) for establishing a program to assess genetically engineered organisms before they are released into the environment. Three commenters — a member of Congress, a spokesperson for a biotechnology firm, and an officer of an environmental organization — posed questions about the proposal. The questions concerned the NBIAP's statutory or regulatory status; its relation to other USDA agency operations and other Federal agency operations; the processes of risk assessment to be used; its adequacy to review an increasing volume of products; and the appropriateness of biohazard committees as vehicles for review of commercial processes and products. Four respondents representing NASULGC institutions endorsed the proposal stating the view that the agricultural research community has the capability to develop guidelines and assess impacts of biotechnology research and commercial products. The major goal of the program was thought to be insuring the safety of society and the environment.

Response: NBIAP is a scientific advisory system that would be available to the Assistant Secretary for Science and Education. By this system the USDA can draw upon the best experience available from scientists in universities, Federal laboratories, and industry to help assess the risks involved in the processes and products from RDNA work in biotechnology.

NBIAP shall act in an advisory capacity and is in no direct way a part of the formal approval process. It is available to provide assessment, but is not a mandatory process.

Comments on Definitions, Terms, and Data Requirements: Five respondents recommended changes in the definitions, terms, data requirements or classification used by USDA in the notice. Each recommendation is discussed below.

Two respondents commented on the USDA statement of licensing policy for veterinary biologics produced by modern biotechnical methods at 49 FR 50899-50900. Under the heading "1. Recombinant DNA-Derived Products," a manufacturer of veterinary biologics questioned the need to provide the entire nucleotide sequence of a foreign DNA being cloned into a vector.

It is USDA's position that in order to characterize adequately the foreign DNA used to code a particular antigen, the manufacturer should provide a nucleotide sequence analysis. The construction of the vector used for expression of the cloned nucleotide sequence also should include source and function of the component parts of the vector, i.e., origin of replication, antibiotic resistance genes, promotor, enhancers, etc. The manufacturer also questioned the data requirement under the heading "2. Chemically Synthesized Antigens" concerning the persistence of the immune response following administration of the synthetic peptide. The USDA feels that a major concern with the use of synthetic peptides is the development persistence of the immune response. USDA does not intend to require more stringent efficacy data than that necessary to support a veterinary biologic license application employing natural antigens. However, immunological data derived from chemically synthesized peptides must be as definitive as the serological response from natural or nonsynthetic antigens. With respect to the next sentence in the policy statement, an individual respondent proposed a change from the term "antibody response" to "immune response." It is true that the term used in the sentence "Procedures used to increase or prolong an antibody response . . ." is somewhat limiting and can create confusion between B-cell and T-cell response. Therefore, the recommendation to replace "antibody response" with the term "immune response" with the term "immune response" is accepted, since both T-cell responses as well as T-cell/B-cell interactions would be included in the statement.

On the subject of plants and plant pests, a plant pathologist commented on the references to *Pseudomonas syringae* as plant pathogens under the heading "ice nucleation negative bacteria" at 49 FR 50902. The respondent noted that none of the strains of *Pseudomonas syringae* currently proposed for use are plant pathogens and that it would be more correct to call *P. syringae* plant-associated bacteria, some of which are pathogens. USDA will clarify future references to these organisms as the respondent suggests. According to current practice, and under the proposed FPPA regulations, an applicant for a USDA permit to import or move *Pseudomonas syringae* would be required to submit data to show whether of not the strain was a plant pest.

Addendum — Research Legislative Authorities

The USDA is authorized under its Organic Act (7 U.S.C. 2201 et seq.) and other legislation to conduct and support research in wide ranging areas of agriculture. Examples of such other laws include:

The Alcohol Fuels Research (7 U.S.C. 3154); the National Latex Commercialization and and Economic Development Act (7 U.S.C. 178-178n); the Animal Health and Disease Research Act (7 U.S.C. 3195); Special Research Grants (7 U.S.C. 450i(c)); The National Aquaculture Act (16 U.S.C. 2801 et seq.); the Cotton Research and Promotion Act (7 U.S.C. 2101 et seq.); the Potato Research Information Act (7 U.S.C. 2611-2627); the Egg Research and Consumer Information Act (7 U.S.C. 701 et seq.); the Beef Research and Information Act (7 U.S.C. 2901 et seq.); the Wheat and Wheat Foods Research and Nutrition Education Act (7 U.S.C. 3401 et seq.); the Animal Cancer Research Act (7 U.S.C. 3901 et seq.); the Floral Research and Consumer

Information Act (7 U.S.C. 4301 et seq.); and the Forest Research Assistance Act (16 U.S.C. 582a-582a-7).

AGENCY GUIDELINES ON BIOTECHNOLOGY

A. Background

The Occupational Safety and Health Administration (OSHA) published an announcement of guidelines on occupational safety and health in the Federal Register, Volume 50, Number 71, page 14483, April 12, 1985 with a request for public comment. All comments received supported the statement although the National Institute for Occupational Safety and Health (NIOSH) recommended increased surveillance because their research indicated gaps in current knowledge.

Biotechnology is the application of biological systems and organisms to technical and industrial processes. The technologies employed in this area include, but are not limited to:

(1) Classical genetic selection and/or breeding for purposes such as developing bakers yeast, conventional fermentation and vaccine development;

(2) The direct in vitro modification of genetic material, e.g., recombinant DA or gene splicing; and,

(3) Other novel techniques for modifying genetic material of living organisms, e.g., cell fusion and hybridoma technology.

Modern biotechnology is analogous to other conventional industrial processes and has great potential benefit to society and wide application to numerous industries. It is considered by some to have economic potential comparable to the microprocessor industry. Genetic engineering has a wide spectrum of applications of commercial importance, but many such applications are in the early stages of development or have been expressed only as concepts.

The Occupational Safety and Health Act of 1970 (OSH Acr) grants the Secretary of Labor broad power to require employers to provide a safe and healthful workplace for their employees. Where other Federal agencies exercise their statutory authority to prescribe or enforce standards or regulations affecting occupational safety or health, OSHA is preempted by section 4(b)(1) of the Act.

Section 5(a)(1) of the Act requires employers to furnish their employees with a workplace "free from recognized hazards that are causing or are likely to cause death or serious physical harm." Section 5(a)(2) requires employers to comply with safety and health standards set by the Secretary. The Secretary in establishing standards to deal with toxic materials and harmful physical agents is required by the OSH Act to "set the standard which most adequately assures, to the extent feasible, on the basis of the best available evidence, that no employee will suffer material impairment of health or functional capacity even if such employee has regular exposure to the hazard dealt with by such standard for the period of his working life" (section 8(b)(5)). Under a recent Supreme Court decision permanent standards can be promulgated only upon a finding by the Secretary that the standard is reasonably necessary to remedy a significant risk of material health impairment. Finally, emergency temporary standards may be promulgated only upon a finding that employees are "exposed to grave danger." (section 8(c)(1).)

In view of the statutory criteria briefly outlined above and the currently known hazards from biotechnology processes there does not appear to be a need for new OSHA regulations. Furthermore, the biotechnology processes, whether present in laboratories, pilot projects or industrial plants, usually involve conventional chemicals and processes that are already covered by OSHA regulations. These conventional processes use solvents or products, some of which may be toxic or dangerous to employee health in certain dosages over certain periods of time. The potentially hazardous character of some aspects of biotechnology is primarily from the chemicals used and not the biotechnology products. Therefore, the regulations that effectively regulate chemical exposures will usually ensure that biohazards too will be controlled. However, when a process employing biotechnology alone or in combination with conventional chemicals and technology presents a significant hazard to employees which cannot be dealt with by existing standards or the general duty clause, OSHA will consider regulating in order to protect employees health. Increased industrial hygiene monitoring and medical surveillance will help to assure worker protection. At this time, no new regulations that would specifically cover biohazards are warranted.

OSHA endorses the BSCC definitions of "intergeneric (new) organism" and "pathogen" found in the preamble, believing they describe the microorganism appropriate for review when environmental or agricultural applications of microorganisms are contemplated. For contained commercial manufacturing processes, these definitions may also properly exclude from review certain microorganisms of known low risk.

OSHA is committed to the policy described in the section entitled "International Aspects" in the Office of Science and Technology Policy General Preamble, published in today's Federal Register.

B. Guidelines

As stated above, section 5(a) of the OSH Act requires

that each employer:

(1) Shall furnish to each of his employees employment and a place of employment which are free from recognized hazards that are causing or are likely to cause death or serious physical harm to his employees;

(2) Shall comply with occupational safety and health standards under this Act.

Specific standards which may be applicable include:

— Specific air contaminants (29 CFR Part 1910, Subpart Z).
— Access to employee exposure and medical records (29 CFR 1910.20).
— Hazard communication (29 CFR 1910.1200).
— Exposure to toxic chemicals in laboratories (currently in draft and under development).
— Respiratory protection (29 CFR 1910.134) (currently being updated).
— Safety standards of a general nature, for example, general environmental, walking and working surfaces, fire protection, compressed gases, electrical safety, and material handling and storage contained in 29 CFR Part 1910 Subparts J, D, E and L, H, S and N).

Effective biological safety and health programs have been operative in a variety of laboratories for many years. Motivation and critical judgment are necessary in addition to specific safety and health knowledge to ensure protection of personnel, the public and the environment. All personnel directly involved in biotechnological projects should receive adequate instruction so that the potential biohazards can be understood and appreciated. Emergency plans should be formulated for each project where the chemicals used or biotechnical product product produced pose a potential safety or health hazard. The plans should describe the procedures to be followed if an accident contaminates personnel or workplaces. If a research group is working with a known pathogen for which an effective vaccine is available, employees should be immunized, as appropriate.

Before biotechnological work is undertaken, it is imporant that management determine the potential hazards involved and the precautions to be taken. Program and support staff should then be advised of the real and potential hazards. Staff should be instructed and trained in the protection and techniques required to ensure safety and in the procedures for dealing with accidentally created hazards.

STATEMENT OF POLICY FOR REGULATING BIOTECHNOLOGY PRODUCTS

In the Federal Register of December 31, 1984 (43 FR 50878), the Food and Drug Administration (FDA) published a policy statement for regulating biotechnology products. The policy statement was part of a larger document that included an index of U.S. laws related to biotechnology, a description of the policies of the major regulatory agencies that are involved in reviewing the products of biotechnology, a description of a proposed scientific advisory mechnism for assessment of biotechnology issues, and an explanation of how the activities of the Federal agencies involving biotechnology will be coordinated. Of the comments FDA received on the policy statement, most favored the policy statement; some requested further clarification and guidance. The current action constitutes FDA's final policy statement which has been revised in response to the comments.

FDA's policy statement of December 31, 1984 stated the FDA regulation must be based on the rational and scientific evaluation of products, and not on *a priori* assumptions about certain processes. Accordingly, FDA's administrative review of products, including those that employ specialized biotechnological techniques, is conducted in the light of the intended use of a product on a case-by-case basis. FDA believes the agency need not establish new administrative procedures to deal with generic concerns about biotechnology.

These views were supported by the majority of comments received in response to FDA's notice. Thirty-four comments were received, with 12 from manufacturers of regulated products, 16 from associations and universities, and 6 from individuals. A summary of the comments and the agency's response to them follow:

1. Many commenters urged the agency to publish additional "Points to Consider" documents to provide further guidance for biotechnology product applicants. These commenters specifically requested guidance in the area of animal drugs (especially protein drugs) and human foods and food additives.

FDA agrees that "Points to Consider" documents provide useful guidance, especially in areas involving new biotechnology, and will consider developing these documents where appropriate.

2. Related comments raised questions on FDA's general requirements for approving bioltechnology products that are animal drugs, human foods, or food additives.

In response to these comments, FDA has amended the animal drug section ("General Requirements for Animal Food Additives and Drugs") to be more informative and has added a new section concerning its policies on human foods and food additives

3. Many comments questioned the need for new or supplemental marketing applications for biotechnology products that are identical to products derived from conventional technology.

The agency has re-examined this issue and continues to believe that, as a general principle, new marketing applications will be required for most products manufactured using new biotechnology. For example, use of recombinant DNA (rDNA) technology has the potential to lead to new structural features in the product, result in product micro-heterogeneity, or introduce new contaminants (e.g., associated with new cell substrates), each of which may affect the safety, efficacy and stability of the product. Because of potential differences in the products resulting from use of recombinant DNA technology, the resulting products may be "new" products requiring separate approval under the applicable statutory provisions. However, each case will be examined separately to determine the appropriate information to be submitted. In some instances complete new applications may not be required. For example, the sponsor of a conventionally produced animal drug product who manufactures an identical or virtually identical product using biotechnology may be required to submit only a supplemental application. However, if the animal drug product manufactured using biotechnology differs significantlyfrom the product manufactured by conventional processes, a complete original application would be required. The agency believes that each product must undergo adequate and appropriate testing and review to ensure that it is safe and effective regardless of the technology employed. Sponsors are urged to communicate with FDA to establish the scope of information required for products of biotechnology.

4. Many comments questioned the need for the proposed review mechanism by a Biotechnology Science Board (BSB).

These comments stated that the additional layer of review would cause delays in the product approval process.

A notice published in the Federal Register of November 14, 1985 (50 FR 47174) discussed the establishment of the Biotechnology Science Coordinating Committee (BSCC) within the Federal Coordinating Council for Science, Engineering and Technology. That notice addressed various criticisms of the BSB. FDA believes that the new BSCC will facilitate sharing of biotechnology information among agencies and will not delay agency reviews of product applications.

In view of the foregoing, FDA's final policy statement for regulating biotechnology products reads as follows:

Introduction

A small but important and expanding fraction of the products the Food and Drug Administration (FDA) regulates represents the fruits of new technological achievements. These achievements are in areas as diverse as polymer chemistry, molecular biology, and micro-miniaturization. It is also noteworthy that technological advancement in a given area may give rise to very diverse product classes, some or all of which may be under FDA's regulatory jurisdiction. For example, new developments in recombinant DNA research can yield products as diverse as food additives, drugs, biologics, and medical devices.

Although there are no statutory provisions or regulations that address biotechnology specifically, the laws and regulations under which the agency approves products place the burden of proof of safety as well as effectiveness of products on the manufacturer. The agency possesses extensive experience with these regulatory mechanisms and applies them to the products of biotechnological processes.

In this notice, FDA proposes no new procedures or requirements for regulated industry or individuals. Rather, the administrative review of products using biotechnology is based on the intended use of each product on a case-by-case basis.

The marketing of new drugs and biologics n1 for human use, and new animal drugs, requires prior approval of an appropriate new drug application (NDA), biological product license, or new animal drug application (NADA). For new medical devices, including diagnostic devices for human use, either a premarket approval application (PMA) or reclassification petition is required. If the device is determined to be substantially equivalent to an already marketed device, a premarket notification under section 510(k) of the Federal Food, Drug, and Cosmetic Act (the act) is required. For food products, section 409 of the act requires preclearance of food additives including those propeared using biotechnology. Section 706 of the act requires preclearance of color additives. The implementing regulations for food and color additive petitions and for affirming generally recognized as safe (GRAS) food substances are sufficiently comprehensive to apply to those involving new biotechnology.

n1 FDA endorises the BSCC definitions of "intergeneric" (new) organism or "pathogen" found in the preamble, believing that they describe the microorganisms appropriate for review when environmental or agricultural applications of the microorganisms are contemplated (and see pp. 22-25). As discussed below in this notice, "new" drugs, biologics, medical devices, and food additives are defined in the statutes establishing FDA's jurisdiction over such products.

Genetic manipulations of plants or animals may enter FDA's jurisdiction in other ways; for example, the introduction into a plant of a gene coding for a pesticide or growth factor may constitute adulteration of foodstuff derived from the plant, or the use of a new microorganism found in a food such as yogurt could be considered a food additive. Such situations will be evaluated case-by-case and in cooperation with the U.S. Department of Agriculture (USDA), where appropriate.

The Regulatory Process

Congress has provided FDA authority under the act and the Public Health Service (PHS) Act to regulate products regardless of how they are manufactured. Each request for product approval will be considered using the appropriate statutory and regulatory criteria. The following sections summarize general requirements for various kinds of products and address specific comments concerning particular product categories. Individual regulations should be consulted for additional details.

General Requirements for New Drugs and Biologics for Human Use

A new drug is, in general terms, a drug not generally recognized by qualified scientific experts as safe and effective for the proposed use. New drugs may not be marketed unless they have been approved as safe and effective for their intended uses. Clinical investigations on human subjects by qualified experts are a prerequisite for the determination of safety and effectiveness. Sponsors of investigations of new drugs or new uses of approved drugs file a Notice of Claimed Investigational Exemption for a New Drug (IND) to conduct clinical investigations on human subjects. The IND must contain information to demonstrate the safety of proceeding to test the drug in human subjects, including, for example, drug composition, manufacturing and controls data, results of animal testing, training and experience of investigators, and a plan for clinical investigation. In addition, assurance of informed consent and protection of the rights and safety of human subjects is required. FDA evaluates IND submissions and reviews ongoing clinical investigations. Significant changes in the conditions of the study, including changes in study design, drug manufacture or formulation, or proposals for additional studies, must be submitted to FDA as amendments to the IND.

FDA approval of an NDA or an abbreviated New Drug Application (ANDA) is required before the new drug can be marketed. The NDA must contain, among other information, the following:

- A list of components of the drug and a statement of the composition of the drug product;
- A description of the manufacturing and packaging procedures and controls for the drug product;
- A description of the nonclinical studies concerning the drug's pharmacological actions and toxicological effects;
- A description and analysis of each clinical study; and
- A description and analysis of any other data or information relevant to an evaluation of the safety and effectiveness of the drug product, including commercial marketing experience.

NDA holders who intend to market an approved drug under conditions other than those approved in the NDA must submit a supplemental NDA containing clinical evidence of the drug's safety and effectiveness for the added indications. Extensive changes such as a change in formula, manufacturing process, or method of testing differing from the conditions of approval outlined in the NDA may also require additional clinical testing.

Biological products must also be approved by FDA prior to marketing, as required by section 351 of the PHS Act. A biological product is "any virus, therapeutic serum, toxin, antitoxin, vaccine, blood, blood component or derivative, allergenic product, or analogous product applicable to the prevention, treatment, or cure of diseases or injuries of man." Unapproved biological products are regulated under the same regulations as new drugs during the IND phase. Prior to marketing, separate licenses are issued for the manufacturing establishment and the biological product. The manufacturing establishment and the biological product must meet standards (including any FDA standards specific for the product) designed to ensure the safety, purity, potency, and efficacy of the product. To obtain a license, the facility must also pass a prelicensing inspection. Licensed products are subject to specific requirements for lot release of FDA.

Manufacturers of new drugs and biologics must operate in conformance with current good manufacturing practice (CGMP) regulations. These regulations require adequately equipped manufacturing facilities, adequately trained personnel, stringent control over the manufacturing process, and appropriate finished product examination. CGMP's are designed to protect the integrity and purity of the product.

The sponsor's process techniques are also considered in FDA's reviews and communications for the development of appropriate information on which the submission of an NDA, ANDA, or biological product license application would be based. For example, the use of recombinant DNA technology to manufacture new drugs or biological products may result in products that differ from similar products manufactured with conventional methods. Determination of the extent of testing required will depend upon the nature of the particular product. In some instances the molecular structure of the product may differ from the structure of the active molecule in nature. For example, the first human growth hormone manufactured using recombinant microorganisms has an extra amino acid, an amino-terminal methionine; hence, it is an analogue of the native hormone. Such differences could affect the drugs's activity or immunogenicity and, consequently, could affect the extent of testing required.

Another consideration in the review of new drugs or biological products producted by recombinant techniques is wheter the manufacturing process includes adequate quality controls. For example, the occurrence of mutations in the coding sequence of the cloned gene during fermentation could give rise to a subpopulation of molecules with an anomalous primary structure and altered activity. This is a potential problem inherent in the production of polypeptides in any fermentation process. As with conventionally produced products, assurance of adequate processing techniques and controls is important in the manufacturing of any biotechnology-produced new drug or biological product. Review of the production of human viral vacciness routinely involves a number of considerations including the purity of the media and the serum used to grow the cell substrate, the nature of the cell substrate, and the characterization of the virus. In the case of live viral vaccine, the final product is biologically active and is intended to replicate in the recipient. Therefore, the composition, concentration, subtype, immunogenicity, reactivity, and nonpathogenicity of the vaccine preparation are all considerations in the final review, whatever the techniques employed in "engineering" the virus. However, special considerations may arise based upon the specific technology employed. For example, a hepatitis B vaccine produced in yeast (via recombinant DNA techniques) would be monitored for yeast cell contaminants, while distinctly different contaminants would be of concern in a similar vaccine produced from the plasma of infected patients.

Nucleic acids or viruses used for human gene therapy will be subject to the same requirements as other biological drugs. It is possible that scientific reviews of these products will also be performed by the National Institutes of Health.

To provide guidance to current or prospective manufacturers of drugs and biological products, the FDA has developed a series of documents describing points that manufacturers might wish to consider in the production and testing of products. The "Points to Consider" documents generated to date include several topics: interferon, monoclonal antibodies, products of recombinant DNA technology, and the use of new cell substrates. These "Points to Consider" documents are available from the agency upon request from the Office of Biological Investigational New Drugs (301-443-4864). FDA plans to develop additional "Points to Consider" in areas of scientific interest to manufacturers of new drugs and biologics.

General Requirements for Animal Food Additives and Drugs

Animal food additives and drugs are subject to similar mandatory requirements of the act as the like products for use in humans. Animal biologics, however, are licensed by the U.S. Department of Agriculture (USDA) under the authority of the Virus-Serum-Toxin Act of 1913. Questions as to whether a product is an animal biological subject to USDA licensure, or a new animal drug to be regulated by FDA are referred to a standing committee of representatives from USDA and FDA.

New animal drugs must go through the Investigational

New animal Drug (INAD) and New Animal Drug Application (NADA) process, a procedure similar to that required for human drugs, as discussed earlier. However, INAD regulations do not require advance agency approval for clinical investigations for the drug, although authorization is required for use of edible products derived from food-producing animals in which the drug has been used. The data must be specific for each animal species for which the drug is intended. For NADA approval, it must be shown that the product is safe and effective when used in accordance with approved label directions. Also, it must be shown that those drugs which are intended for use in food-producing animals and used in accordance with approved label directions, do not accumulate as unsafe residues in the edible tissues of the animal at the time of slaughter. Moreover, the manufacturer must submit acceptable methods for measurement of any drug residue in edible tissues. Further, animal drugs, including premixes for use in medicated feeds and medicated feeds, must be manufactured in conformance with CGMPs. Substances that are used in animal feeds, other than drugs, and that are produced by recombinant DNA technology, are considered to be food additives and require approval of a separate food additive petition (FAP), even though a similar substance is currently approved as a food additive.

There have been questions about the requirement of an orginal application for a biotechnology product, even when the product is identical to a currently approved animal drug held by the same applicant. FDA's Center for Veterinary Medicine (CVM) has determined that, when the new substance produced by biotechnology is identical or virtually identical to an approved substance produced by conventional technology, only a supplemental application is necessary. Of course, in this instance the sponsor of the biotechnology product must also be the sponsor of the conventionally produced product. If, on the other hand, the new substance produced by biotechnology is significantly different from that produced by conventional means, an original application will be needed.

Two examples, each involving the adoption of rDNA technology as an alternative means of producing a substance that is currently the subject of an approved NADA, will illustrate. In the first example, the drug is (or appears to be) unchanged by the new production method. Under the current regulations, such a departure in manufacturing procedure requires a supplemental application which requires approval before implementation. The supplement would be a Category II supplement under CVM's supplemental policy in that it involves a revised method of synthesis or fermentation for the new drug substance. However, in accordance with the CVM's supplemental policy the underlying safety and effectiveness data supporting the original NADA usually would not be reviewed (for compliance with contemporary standards) since there is likely no increased risk of human exposure to the drug. Data may be required to demonstrate the new animal drug product is essentially biologically equivalent to the drug product for which approval has already been granted. Approval of such a supplemental NADA is not required to be published in the Federal Register.

In the second example, a new method of manufacture changes the molecular structure or chemcial composition of the active ingredient. Such a change in the identity of the new animal drug normally wil require an original new animal drug application and subsequent publication of a notice of approval in the Federal Register. Ordinarily, an original NADA requires complete safety and effectiveness studies, meeting contemporary standards. However, reference to data in another NADA sometimes suffices to support a separate NADA approval, where the existing NADA is owned by the applicant of the new NADA, or where the new applicant obtains authorization to reference might be made to data contained in the NADA supporting approval of the drug as produced by conventional means.

It may be possible to regard the new application as if it were a Category II supplement. This finding would be dependent upon data showing the new substance to be sufficiently similar to the original in terms of its pharmacology, toxicology, bioequivalence, and metabolism.

Thus, regardless of the type of application required, there is no legal requirement for the generation of new safety and effectiveness data if the applicant has access to previously submitted data, and there is no scientific need.

General Requirements for Medical Devices

Medical devices for human use are regulated by requirements of the act as amended by the Medical Device Amendments of 1976. In general, a device is a health care product that does not achieve any of its principal intended purposes by chemical action in or on the body or by being metabolized. Devices include diagnostic aids such as reagents, antibiotic sensitivity discs, and test kits for *in vitro* diagnosis of disease.

The act establishes three classes of devices: Class I (general controls), class II (performance standards), and class III (premarket approval). Classification of a device is determined by the level of regulatory control needed to provide reasonable assurance of the safety and effectiveness of the device. A class I device is a device for which the "general controls" authorized by or under various sections of the act are sufficient to provide reasonable assurance of the safety and effectiveness of a device. A class II device is a device for which general controls by themselves are insufficient to provide reasonable assurance of the Safety and effectiveness of the device, for which there is sufficient information to establish a performance standard to provide such assurance, and for which it is therefore necessary to establish a performance standard to provide reasonable assurance of its safety and effectiveness. A class III device is a device that cannot be classified into class I or class II and that is purported or represented to be for use in supporting or sustaining human life or for a use which is of substantial importance in preventing impairment of human, health, or that presents a potential unreasonable risk of illness or injury. Premarket approval obtained in accordance with section 515 of the act is required to provide reasonable assurance of the safety and effectiveness of a class III device.

Before a manufacturer may introduce into commerce any medical device it has not previously marketed, the manufacturer must submit to FDA a premarket notification. This notification requirement is designed to assure that manufacturers do not intentionally or unintentionally circumvent the automatic classification into class III of devices not on the market prior to enactment of the Medical Device Amendments and not substantially equivalent to pre amendment devices.

A new device, that, is one not substantially equivalent to a preamendments device, remains a class III device requiring FDA approval of a premarket approval application (PMA) unless FDA reclassifies it into class I or class II, usually in response to a manufacturer's petition. In the premarket approval process the manufacturer must establish by valid scientific evidence that the device is safe and effective for its intended use. This evidence usually is data from clinical investigations.

For a significant risk device, as defined in FDA's regulations, the sponsor must submit an application to FDA for approval to conduct a clinical investigation. This application seeks an Investigational Device Exemption. When the manufacturer believes that there are sufficient data to establish the safety and effectiveness of its device, the manufacuter files a PMA.

General Requirements for Foods

several sections of the Food, Drug and Cosmetic Act apply to the Agency's regulation of food. No particular statutory provision or regulation deals expressly with food produced by new biotechnology. Accordingly, when confronted by an issue concerning the regulation of food produced by new biotechnology, the Agency will apply the relevant statutory or regulatory provisions. Most issues concerning the safety of a food will involve the application of either section 402(a)(1) or section 409 of the Act.

Section 402(a)(1) of the Act provides, in part, that a food is adulterated if it bears or contains any poisonous or deleterious "added substance" which may render it injurious to health." Courts have agreed with the agency's interpretation of this section that any substance that is not an inherent constituent of food may be regulated as an "added substance." *See,* for example, *United States* v. *Cartons of Swordfish,* 395 F. Supp. 1194 (S.D.N.Y. 1975). Furthermore, if the quantity of the constituent exceeds the amount that would normally be present because of some technological adjustment to the product, that excess quantity may also be viewed as "added substance" within the meaning of the section. *See United States* v. *Anderson Sea Foods, Inc.,* 622 F.2d 157 (5th cir. 1980). Thus, section 401(a)(1) applies to most of the harmful substances that may occur in human food. For example, is a food produced by new biotechnology contains a higher level of a substance than it might ordinarily have, then that level "may be injurious to health" and the agency could regulate the product under section 402(a)(1). Similarly, if a food produced by new biotechnology contains, as a result of the production process, a harmful or deleterious substances not contained ordinarily in the food, the food could be in violation of the section.

The other primary statutory provisions that FDA relies upon in determining the safety of food and food constituents are sections 201(s) and 409, the food additive provisions of the Act. The definition of food additive appears in section 201(s) of the Act and includes both artificial and natural substances. The definition provides that:

the term food additive means any substance the intended use of which results or may reasonably be expected to result, directly or indirectly, in its becoming a component or otherwise affecting the characteristics of any food (including any substance intended for use in producing, manufacturing, packing, processing, preparing, treating, packaging, transporting, or holding food; and including any source of radiation intended for any such use), if such substance is not generally recognized as safe by qualified experts.

If the substance is generally recognized as safe (GRAS) for a given food use, the product is not a food additive.

Comments questioned whether a substance (including microbes) that is GRAS could lose its GRAS status solely because it was produced or modified by new biotechnology. The answer is yes, if the substance (and its contaminants) has been altered in such a way that it can no longer be generally recognized by qualifed experts to be safe. In this instance, the substance would be a food additive and the provisions of section 409 would apply. Section 409 provides that in order to be lawfully used in food, a food additive must be the subject of an approved food additive regulation, published upon approval of a food additive petition. The FDA may not approve a food additive regulation until certain basic evidentiary criteria are met. Most important of these is that the additive must be shown to be safe under the conditions that it will be used. This requires a demonstration to a reasonable certainty that the additive will not adversely affect the health of consumers.

FDA anticipates that the techniques of new biotechnology used in producing food will, for the most part, involve rDNA and microbial isolation. The agency applies certain general principles that it will follow in determining the safety of foods produced by such techniques.

When determining the safety of food produced by rDNA techniques, the agency takes into consideration, but is not restricted to, whether:

1. The cloned DNA as well as the vector used are properly identified;
2. The details of the construction of the production organism are available;
3. There is information documenting that the inserted DNA is well-characterized n2 and free from sequences that code for harmful products, and
4. The food produced is purified, characterized, and standardized.

n2 As defined by the BSCC definitions in the preamble, "well-characterized" means that the producer can document the exact nucleotide sequence of the insert and any flanking nucleotides.

When determining the safety of food produced by microbial isolation, the agency will take into consideration, but is not restricted to, whether:

1. The microbial isolate used for production is identified taxonomically, and if the strain of the isolate has been genetically manipulated, whether each strain contributing genetic information to the production strain is identified;
2. The cultural purity and genetic stability of isolate has been maintained;
3. Fermentation has been performed with a pure culture and monitored for purity;
4. The microbial isolate used for production also produces antibiotics or toxins;
5. The isolates are pathogenic; n3 and
6. Viable cells of the production strain are present in the final product.

n3 A pathogen is a virus or microoganism (including its viruses and plasmids, if any) that has the ability to cause disease in other living organisms (i.e., humans, animals, plants, microorganisms).

A microorganism will be included within this definition if:

a. The microorganism belongs to a pathogenic species, according to sources identified by the agency, or from information known to the producer that the organism is a pathogen; excepted are organisms belonging to a strain used for laboratory research or commercial purposes and generally recognized as nonpathogenic according to sources identified by a federal agency, or information known to the producer and the appropriate federal agency; an example of a nonpathogenic strain of species which contains a pathogenic strain is Escherichia coli K-12; examples of nonpathogenic species are Bacillus subtilis, Lactobacillus acidophilus, and Saccharomyces species; or

b. The microorganism has been derived from a pathogen or has been deliberately engineered such that it contains genetic material from a pathogenic organism as defined in item a above. Excepted are genetically engineered organisms developed by transferring a well-characterized, non-coding regulatory region from a pathogenic donor to a nonpathogenic recipient.

"Well-characterized, non-coding regulatory region" means that the producer of the microorganism can document the following:

a. The exact nucleotide base sequence of the regulatory region and any inserted flanking nucleotides;
b. The regulatory region and any inserted flanking nucleotides do not code independently for protein, peptide, or functional RNA molecules; and,
c. The regulatory region solely controls the activity of other sequences that code for protein or peptide molecules or act as recognition sites for the initiation of nucleic acid or protein synthesis.

This definition excludes organisms such as competitors or colonizers of the same substrates, commensal or mutualistic microorganisms, or opportunistic pathogens.

As a general rule, the extent of testing required on a food product produced by biotechnology will depend upon many factors, including the novelty of the substances used to produce the food (e.g., whether a substance is an "intergeneric" organism, as defined by the BSCC definitions in the preamble), the purity of the resulting product, and the estimated consumption of the product.

The agency will require that the final product intended for commercialization be the article tested. A complete discussion of FDA's toxicology requirements is found in the FDA publication, "Toxicological Principles for the Safety Assessment of Direct Food Additives and Color Additives Used in Food." This publication is available through the National Technical Information Service (publication # PB 83-170696) 5285 Port Royal Road, Springfield, VA 22161. Questions concerning the publication can be directed to Dr. Alan M. Rulis in the Center for Food Safety and Applied Nutrition (CFSAN) at (301) 472-5676.

Obligations Under the National Environmental Policy Act

All premarketing approvals of FDA-regulated products are subject to the requirements of the National Environmental Policy Act (NEPA) as defined by the Council on Environmental Quality's regulations (40 CFR Parts 1500-1508) and as further described by FDA's NEPA-implementing procedures (21 CFR Part 25, final rule published April 26, 1985; 50 FR 16636). For new products or major new uses for existing products, these procedures ordinarily require the preparation of an environmental assessment. An environmental impact statement is required if the manufacture, use, or disposal of the product is anticipated to cause significant environmental impacts.

International Aspects

FDA is committed to the policy described in the section entitled "International Aspects" in the Office of Science and Technology Policy General Preamble, published in today's Federal Register.

REFERENCES

Adachi S, et al. Cyclin A/cdk2 activation is involved in hypoxia-induced apoptosis in cardiomyocytes Circ Res. 88:408, 2001.

Agocha A, Lee H-W, Eghali-Webb M. Hypoxia regulates basal and induced DNA synthesis and collagen type I production in human cardiac fibroblasts: effects of TGF-beta, thyroid hormone, angiotensis II and basic fibroblast growth factor. J Mol Cell Cardiol 1997;29:2233-2244H.

Amit M, Carpenter MK, Inokuma MS, Chiu C, Harris CP, Waknitz MA, Itskovitz-Eldor J, Thomson JA. Clonally derived human embryonic stem cell lines maintain pluripotency and proliferative potential for prolonged periods of in vitro culture. Developmental Biology 2000; 227: 271-278.

Amit, M., et al., "*Clonally de*rived Human Embryonic Stem Cell Lines Maintain Pluripotency and Proliferative Potential for Prolonged Periods of Culture," Developmental Biology 227: 271-278 (2000); Amit, M. a*nd Itskovitz-Eldor, J.*, "Derivation and spontaneous differentiation of human embryonic stem cells," Journal of Anatomy 200: 225-232 (2002).

Andres, V. and Walsh, K. (1996) Myogenin expression, cell cycle withdrawal and phenotypic differentiation are temporally separable events that precedes cell fusion upon myogenesis. J. Cell Biol., 132, 657-666.

Anversa P, and Nadal-Ginard B. Myocyte renewal and ventricular remodelling Nature 415, 240 - 243, 2002.

Auchincloss H Jr (2001) In search of the elusive holy grail: the mechanisms and prospects for achieving clinical transplantation tolerance. Am J Transplant 1:6-12.

Beltrami AP, et al. Evidence that human cardiac myocytes divide after myocardial infarction. *N Engl J Med* 344, 1750-7 (2001).

Brent L (1997) Fetally and neonatally induced immunologic tolerance. In: Brent L (ed) A history of transplantation immunology. Academic, San Diego.

Carpenter MK, Inokuma MS, Denham J, Mujtaba T, Chiu CP, Rao MS.

Carpenter, *M. K., et al.*, "Characterization and Differentiation of Human Embryonic Stem Cells," Cloning and Stem Cells 5: 79-88 (2003).

Drukker M, Katz G, Urbach A, Schuldiner M, Markel G, Itskovitz-Eldor J, Reubinoff B, Mandelboim O, Benvenisty N. Characterization of the expression of MHC proteins in human embryonic stem cells. Proc Natl Acad Sci U S A 2002; 99: 9864-9869.

Eiges R, Schuldiner M, Drukker M, Yanuka O, Itskovitz-Eldor J, Benvenisty N. Establishment of human embryonic stem cell-transfected clones carrying a marker for undifferentiated cells. Current Biology 2001; 11: 514-518.

Goldstein RS, Drukker M, Reubinoff BE, Benvenisty N. Integration and differentiation of human embryonic stem cells transplanted to the chick embryo.

Gronthos, S., *et al.*, "*Molecul*ar and cellular characterization of highly purified stromal stem ceils derived from bone marrow," Journal of Cell Science 116: 1827- 1835 (2003).

Halevy, O., Novitch, B.G., Spicer, D.B., Skapek, S.X., Rhee, J., Hannon, G.J., Beach, D. and Lassar, A.B. (1995) Correlation of terminal cell cycle arrest of skeletal muscle with induction of p21 by MyoD. Science, 267, 1018- 1021.

He JQ, Ma Y, Lee Y, Thomson JA, Kamp TJ. Human Embryonic Stem Cells Develop Into Multiple Types of Cardiac Myocytes. Action Potential Characterization.

Henderson JK, Draper JS, Baillie HS, Fishel S, Thomson JA, Moore H, Andrews PW. Preimplantation human embryos and embryonic stem cells show comparable expression of stage-specific embryonic antigens. Stem Cells 2002; 20: 329-337.

Heymans S, Luutun A, Nuyens D, et al. Inhibition of plasminogen activators or matrix metalloproteinases prevents cardiac rupture but impairs therapeutic angiogenesis and causes cardiac failure. Nat Med 1999;5:1135- 1142.

Hill MF, Singal PK. Right and left myocardial antioxidant responses during heart failure subsequent to myocardial infarction. Circulation 1997 96:2414-20.

Hochman JS and Choo H. Limitation of myocardial infarct expansion by reperfusion independent of myocardial salvage. Circulation 1987;75:299- 306.

Hognes JR. In The Artificial Heart: Prototypes, Policies and Patients. Washingtion, DC: National Academy Press; 1991:1-312.

Janeway CAJ (1992) The immune system evolved to discriminate infectious nonself from noninfectious self. Immunol Today 13:11-16.

Kajstura J, Leri A, Finato N, di Loreto N, Beltramo CA, Anversa P. Myocyte proliferation in end-stage cardiac failure in humans. Proc Natl Acad Sci USA 95, 8801-8805 (1998).

Kalkman EAJ, Bilgin YM, van Haren P, van Suylen R-J, Saxena PR, Schoemaker RG.Determinants of coronary reserve in rats subjected to coronary artery ligation or aortic banding. Cardiovasc Res 1996.

Kehat I, Gepstein A, Spira A, Itskovitz-Eldor J, Gepstein L. High-resolution electrophysiological assessment of human embryonic stem cell-derived cardiomyocytes: a novel in vitro model for the study of conduction. Circ Res 2002; 91: 659-661.

Kehat I, Kenyagin-Karsenti D, Snir M, Segev H, Amit M, Gepstein A, Livne E, Binah O, Itskovitz-Eldor J, Gepstein L. Human embryonic stem cells can differentiate into myocytes with structural and functional properties of cardiomyocytes. J Clin Invest 2001; 108: 407-414.

Koc, O. N., et al., "*Rapid he*matopoietic recovery after coinfusion of autologousblood stem cells and culture-expanded marrow mesenchymal cells in advanced breast cancer patients receiving high-dose chemotherapy," Journal of Clinical Oncology 18: 307-316 (2000).

Kondo, M., *et al.*, "*Biology of H*ematopoietic Stem Cells and Progenitors: Implications for Clinical Application," Annual Review of Immunology 21: 759-806 (2003) and re*ferences cited therein.*

Krampera M, et al (2002). Bone marrow mesenchymal stem cells inhibit the response of naive and memory antigen-specific T cells to their cognate peptide.

Lanzendorf SE, Boyd CA, Wright DL, Muasher S, Oehninger S. Use of human gametes obtained from anonymous donors for the production of human embryonic stem cell lines. Fertility and Sterility 2001; 76: 132-137.

Lebkowski JS, Gold J, Xu C, Funk W, Chiu CP, Carpenter MK. Human embryonic stem cells: culture, differentiation, and genetic modification for regenerative medicine applications. Cancer J 2001; 7: S83-93.

Levkau B, Koyama H, Raines EW, Clurman BE, Herren B, Orth K, Roberts JM, Ross R. Cleavage of p21cip1/waf1 and p27 kip1 mediates apoptosis in endothelial cells through activation of cdk2: role of a caspase cascade.

Li, Y., Jenkins, C.W., Nichols, M.A. and Xiong, Y. (1994) Cell cycle expression and p53 regulation of the cyclin-dependent kinase inhibitor p21. Oncogene, 9, 2261-2268 **24.** Steinman, R.A., Hoffman, B., Iro, A., Guillouf, C., Liebermann, D.A. and El-Houseini, M.E. (1994) Induction of p21 (WAF1/CIP1) during differentiation.

Liechty KW, et al (2000). Human mesenchymal stem cells engraft and demonstrate site-specific differentiation after in utero transplantation in sheep.

Lodie, T. A., et *al.*, "*Systematic* analysis of reportedly distinct populations of multipotent bone marrow-derived stem cells reveals a lack of distinction," Tissue Engineering 8: 739-751 (2002).

Ma Y, Ramezani A, Lewis R, Hawley R, Thomson JA. High-level sustained transgene expression in human embryonic stem cells using lentiviral vectors. Stem Cells 2003; 21: 111-117.

MacLellan WR and Schneider MD. Genetic dissection of cardiac growth control pathways Annu. Rev. Physiol. 2000. 62:289-320.

Mahon NG, O'Roke C, Codd MB, et al. Hospital mortality of acute myocardial infarction in the thrombolytic era. Heart 1999;81:478-82.

Makino S, Fukuda K, Miyoshi S, et al. Cardiomyocytes can be generated from marrow stromal cells in vitro. J Clin Invest 1999, 103:697-705.

Matzinger P (1994) Tolerance, danger, and the extended family. Annu Rev Immunol 12:991-1045.

Mellor AL, Munn DH (2001) Extinguishing maternal immune responses during pregnancy: implications for immunosuppression. Semin Immunol 13:213-218.

Moore FL, Jaruzelska J, Fox MS, Urano J, Firpo MT, Turek PJ, Dorfman DM, Pera RA. Human Pumilio-2 is expressed in embryonic stem cells and germ cells and interacts with DAZ (Deleted in AZoospermia) and DAZ-like proteins. Proceedings of the National Academy of Sciences of the United States of America 2003; 100: 538- 543.

Mummery C, Ward-Van Oostwaard D, Doevendans P, Spijker R, Van Den Brink S, Hassink R, Van Der Heyden M, Opthof T, Pera M, De La Riviere AB, Passier R, Tertoolen L. Differentiation of human embryonic stem cells to cardiomyocytes: role of coculture with visceral endoderm-like cells. Circulation 2003; 107: 2733-2740.

Nelissen-Vrancken H, Debets J, Snoeckx L, Daemen M, Smits J. Time-related normalization of maximal coronary flow in isolated perfused hearts of rats with myocardial infarction. Circulation 1996;93:349-355.

Nidorf SM, Siu SC, Galambos G, Weyman AE, Picard MH. Benefit of late coronary reperfusion on ventricular morphology and function after myocardial infarction. J Am Coll Cardiol 1992;20:307-313.

Pfeffer JM, Pfeffer MA, Fletcher PJ, Braunwald E. Progressive ventricular remodeling in rat with myocardial infarction. Am J Physiol 1991; 260:H1406-14.

Pfeifer A, Ikawa M, Dayn Y, Verma I. Transgenesis by lentiviral vectors: Lack of gene silencing in mammalian embryonic stem cells and preimplantation embryos. PNAS 2002; 99: 2140-2145.

Pittenger, *M. F. et al., "Multil*ineage potential of adult human mesenchymal stem cells," Science 284: 143-147 (1999); Pittenger, M., et al., "*Adult me*senchymal stem cells: Potential for muscle and tendon regeneration and use in gene therapy," Journal of Musculoskeletal and Neuronal Interactions *2: 309-320 (2002).*

Reubinoff BE, Itsykson P, Turetsky T, Pera MF, Reinhartz E, Itzik A, Ben- Hur T. Neural progenitors from human embryonic stem cells. Nat Biotechnol 2001; 19: 1134-1140.

Reubinoff BE, Pera MF, Vajta G, Trounson AO. Effective cryopreservation of human embryonic stem cells by the open pulled straw vitrification method. Hum Reprod 2001; 16: 2187-2194.

Sathananthan H, Pera M, Trounson A. The fine structure of human embryonic stem cells. Reprod Biomed Online 2002; 4: 56-61.

Schuldiner M, Eiges R, Eden A, Yanuka O, Itskovitz-Eldor J, Goldstein RS, Benvenisty N. Induced neuronal differentiation of human embryonic stem cells. Brain Research 2001; 913: 201-205.

Schuldiner M, Yanuka O, Itskovitz-Eldor J, Melton DA, Benvenisty N. Effects of eight growth factors on the differentiation of cells derived from human embryonic stem cells. Proc Natl Acad Sci U S A 2000; 97: 11307-11312.

Soonpaa MH, Field LJ. Assessment of cardiomyocyte DNA synthesis in normal and injured adult mouse hearts. *Am J Physiol* **272**, H220-6 (1997).

Steinberg, D., "Can Moral Worthiness Be Seen Using a Microscope?" *American Journal of Bioethics* 2(1): 49 (2002).

Thomson AW, Lu L (1999) Dendritic cells are regulators of immune reactivity: implications for transplantation. Transplantation 68:1-8.

Tomita S, Li R-K, Weisel RD, et al. Autologous transplantation of bone marrow cells improves damaged heart function. Circulation 1999; 100:II-247.

Wekerle T, Kurtz J, Ito H, Ronquillo JV, Dong V, Zhao G, Shaffer J, Sayegh MH, Sykes M (2000) Allogeneic bone marrow transplantation with co-stimulatory blockade induces macrochimerism and tolerance without cytoreductive host treatment. Nat Med 6:464-469.

White HD, Cross DB, Elliot JM, et al. Long-term prognostic importance of patency of the infarct-related coronary artery after thrombolytic therapy for myocardial infarction. Circulation 1994;89:61-67.

White HD, Norris RM, Brown MA, Brandt PWT, Whitlock RML, Wild CJ. Left ventricular end systolic volume as the major determinant of survival after recovery from myocardial infarction. Circulation 1987; 76:44-51.

Zhang SC, Wernig M, Duncan ID, Brustle O, Thomson JA. In vitro differentiation of transplantable neural precursors from human embryonic stem cells.

Zwaka TP, Thomson JA. Homologous recombination in human embryonic stem cells. Nat Biotechnol 2003; 21: 319-321.

5

Biotechnology and Telecommunication: Modern Innovations

Deciding about the range of possible policy interventions requires a re-assessment of the extent to which emerging technologies are influenced by general principles, as opposed to the vagrancies of historical circumstances. Our ambition with devoting a special issue to modern biotechnology and telecommunication is to encourage others—and ourselves—to move beyond the details of innovation in specific technologies to strive for more systematic and comprehensive comparisons.

Problems to be Addressed

The social scientific literature has long established that innovations based on emerging technologies are crucial for growth. In spite of this, the specific interactions among science and technology that lead to economic growth remain a black box in many ways. In particular, we still have a limited understanding of the conditions and processes that make it possible for actors to innovate, based on new technologies. For instance, there is an abundance of theories concerning the role of intellectual capital in creating value, but relatively little empirical evidence explaining the nature of the linkage between such capital, innovation and commercial success of a firm. Similarly, the concept of technological discontinuities is often applied to emerging technologies, in order to stress the radical nature of certain innovations. Yet, it is not clear why some actors manage to take advantage of technological discontinuities, while others fail to grasp the opportunities associated with such changes and thereby suffer from them. Thus, even though innovations are now widely accepted to contribute positively to firm and economic growth, the specifics at the micro-economic level often tell a story of diversity, experimentation and selection, and in doing so, open up new puzzles for explaining how and why innovation causes growth.

Micro-economic Mechanisms

Existing literature on innovation has made many steps toward sorting out general starting conditions for innovation from historical events. Yet, we would argue that in order to understand economic growth on a macro level, it is crucial to understand the micro-economic mechanisms that drive technological development and innovation. The paradox of setting out to consider the general conditions for innovation is that the micro-economic mechanisms tend to express themselves in a variety of ways. As outlined in the subsequent materials, some of the mechanisms driving innovation consist of experimentation, collaboration and knowledge-creation-yet discovery is at most half the story. Other mechanisms involve commercial exploitation, such as adaptation of knowledge to specific industries or adaptation of services to specific groups of users. It appears that aspects such as the maturity of the industry and nature of the technology involved most likely affect how these micro-economic mechanisms are expressed within a particular case.

Empirical Diversity

Empirical 'reality' about emerging technologies often

provides us with a diverse wealth of empirical details. The difficulty of 'testing' theoretical ideas in a strict sense has to do, we would argue, with the normal course of economic dynamics, where diversity and selection (based on criteria) affect the outcomes. The outcome is not given from the beginning, but instead unfolds through historical processes, driven by individuals as they interpret societal structures. Indeed, empirical reality provides us a great diversity of success, failures and experience -and these results may sometimes support (or reject) predictions from one theory. Yet the empirical diversity surrounding emerging technologies more often provides partial support to several existing, sometimes competing, theories—or they may help develop new ways of conceptualizing the modern economy. Theory-driven empirical studies have been common for research on both modern biotechnology and telecommunication, and lead to new concepts and theories about innovation, including examples of 'knowledge spill-overs', 'network externalities', and 'competition among standards'. The apparent vagueness of human factors affecting innovation should not discourage study of the conditions for innovation, but rather provide motivation for more systematic research efforts.

Human Versus Structural Capital

The broader perspective on modern biotechnology and telecommunication provides many interesting points of comparison. On the one hand, modern biotechnology is a knowledge-based industry that depends on complex, and to some extent ambiguous, research activities. The knowledge underlying innovations in biotechnology is often located within specific individuals and organizations, thereby making difficult to separate those who produce it from those who possess it for commercial gains. Many companies are striving to turn this human capital into structural capital, but, so far, individuals interacting in teams and through social links are responsible for most of the innovative activity resulting in new products and services. As a result, the biotechnology industry provides an excellent opportunity to investigate the relationship between individuals, innovation and commercial success. Telecommunications, on the other hand, is interesting not primarily because of the role of individuals, but as a technological system of many interacting forces and actors. The human capital held by individuals has to a greater extent value within the larger technological system. The ability to innovate thereby often relies on a whole series of services, hardware, development teams, existing computer programs, user behaviour, etc. Even so, both are highly dynamic and complex technologies affecting industries, which are characterized by interaction between technological change, markets, demand, and regulation.

Within this broader perspective, this special issue considers three themes related to modern biotechnology and telecommunication.

Theme 1: Evolutionary, Long Term Processes

One theme is that innovation processes are largely characterized as an evolutionary, long term process. This process involves exploration and exploitation of new knowledge and of new commercial opportunities by a variety of actors. Another way of stating this theme is that the characteristics include diversity, much experimentation, failures and successes, organizational heterogeneity, and the like. While these characteristics apply to many technologies and industrial sectors, they are highly visible within these two highly dynamic and complex technologies. This first theme implies that stimulating innovation characteristics—rather than trying to rationally design outcomes—should be the aim when designing regulation and control mechanisms. For example, deciding upon a centralized, tight control of inputs into innovation processes in order to achieve one pre-defined outcome may not work terribly well. Instead, one must consider how delineating the characteristics can be used to design policies stimulating such processes. In some instances, government and firm policy needs to differentiate the roles of organizations/actors, in order to specify which parts of the exploration and exploitation process each is/should be involved in. Different roles are possible. Therefore, we do not suggest a romantic looking-back to the telecoms era with government sponsored telcos and nationally favored equipment suppliers. Even so, in the present increasingly competitive market place, there is room for government policies with industrial ambitions and goals to stimulate the development of technologies.

Theme 2: Role of Infrastructure

The second theme is that despite diversity and experimentation, both modern biotechnology and telecommunication can be characterized as dependent upon infrastructures. Especially telecommunication—but also biotech—involves many technological, social and organizational components, which together create broader infrastructures, or technological systems. For example, successfully applying biotechnology techniques to identify new molecules useful for human disease treatments requires that this bit of information is assessed relative to a large variety of knowledge, regulation, and industries to become a pharmaceutical product taken by patients. Similarly, in telecommunication, the physical infrastructural investment in communication equipments requires many complex components, but also relies upon consumers' preferences and behavior as well as available services and software. This second theme thereby implies that infrastructural characteristics often help determine the degree of freedom in innovation which is open to the individual actor. For example, the development of a standard may open a series of low-risk business opportunities for small firms. This may stimulate experimentation by a great variety and number of actors—but a standard can lead to dominance by one main actor as well. Another aspect of an infrastructure dependency is that it is difficult to separate competition from cooperation. Instead, the common infrastructure forces market players to cooperate while at the same time competing.

Theme 3: Industries, Firms or Products?

The third theme is that analytical tensions exist between seeing modern biotechnology and telecommunication as industries—as opposed to technologies composed of knowledge, equipment, techniques, etc. On the one hand, much existing literature describes each in terms of industries and industrial sectors, whereby competition is explained in terms of competing products or competing firms. The traditional case is to analyze products and firms as competing on price. On the other hand, both technologies are applicable to a wide range of industries, and they must be further developed in terms of knowledge and the infrastructure to be really applicable there. This case is more relevant when considering competition based not only on price but also on innovation. Some areas of technology will give higher economic return than other areas, leading to rather skewed results in terms of search activities and resulting innovations.

The articles in this special issue strive to shed light on these issues, and pave the way for further research, by presenting empirical evidence and new perspectives on innovation from two emerging technologies—or two dynamic industries. Our ambition here is to fuel debates through some current findings, but also encourage others to assess a number of problem areas associated with the larger question of theoretical abstraction versus historical specificities.

Certain aspects of innovation may be difficult to deconstruct—especially in areas where technological change is rapid—but this does not mean that these structures and conditions that do affect innovation and growth should be ignored. Indeed, the existence of complex and difficult issues suggests the need for systematic analysis as well as suggestions for intervention by public and private actors.

REFERENCES

Abraham, E. J., et al., "Insulinotropic hormone glucagons-like peptide-1 differentiation of human pancreatic islet-derived progenitor cells into insulinproducing cells," Endocrinology 143: 3152-3161 (2002).

Adachi S, et al. Cyclin A/cdk2 activation is involved in hypoxia-induced apoptosis in cardiomyocytes Circ Res. 88:408, 2001.

Agocha A, Lee H-W, Eghali-Webb M. Hypoxia regulates basal and induced DNA synthesis and collagen type I production in human cardiac fibroblasts: effects of TGF-beta, thyroid hormone, angiotensis II and basic fibroblast growth factor. J Mol Cell Cardiol 1997;29:2233-2244H.

Alan Meisel of the University of Pittsburgh School of Law, quoted in Coyle, M., "The Clone Zone," *The National Law Journal*, May 15, 2002.

Alpers, Ann, and Bernard Lo. "Commodification and Commercialization in Human Embryo Research." Stanford Law and Policy Review, vol. 6, no. 2 (1995): 40.

American Association for the Advancement of Science and Institute for Civil Society.

Baker, K. S., et *al., "New* Malignancies After Blood or Marrow Stem-Cell Transplantation in Children and Adults: Incidence and Risk Factors," Journal of Clinical Investigation 21: 1352-1358 (2003).

Banerjee, N. & Zhang,M.X. Functional genomics as applied to mapping transcription regulatory networks.*Curr.Opin.*

Beauchamp JR, Morgan JE, Pagel CN, Partridge TA (1999) Dynamics of myoblast transplantation reveal a discrete minority of precursors with stem celllike properties as the myogenic source. J Cell Biol 144:1113-1122.

Beddington RS, Robertson EJ. An assessment of the developmental potential of embryonic stem cells in the midgestation mouse embryo. Development 1989; 105: 733-737.

Kajstura J, Leri A, Finato N, di Loreto N, Beltramo CA, Anversa P. Myocyte proliferation in end-stage cardiac failure in humans. Proc Natl Acad Sci USA 95, 8801-8805 (1998).

Kalkman EAJ, Bilgin YM, van Haren P, van Suylen R-J, Saxena PR, Schoemaker RG.Determinants of coronary reserve in rats subjected to coronary artery ligation or aortic banding. Cardiovasc Res 1996.

Kehat I, Gepstein A, Spira A, Itskovitz-Eldor J, Gepstein L. High-resolution electrophysiological assessment of human embryonic stem cell-derived cardiomyocytes: a novel in vitro model for the study of conduction. Circ Res 2002; 91: 659-661.

Kehat I, Kenyagin-Karsenti D, Snir M, Segev H, Amit M, Gepstein A, Livne E, Binah O, Itskovitz-Eldor J, Gepstein L. Human embryonic stem cells can differentiate into myocytes with structural and functional properties of cardiomyocytes. J Clin Invest 2001; 108: 407-414.

Koc, O. N., et al., *"Rapid he*matopoietic recovery after coinfusion of autologousblood stem cells and culture-expanded marrow mesenchymal cells in advanced breast cancer patients receiving high-dose chemotherapy," Journal of Clinical Oncology 18: 307-316 (2000).

Kondo, M., *et al., "Biology of H*ematopoietic Stem Cells and Progenitors: Implications for Clinical Application," Annual Review of Immunology 21: 759-806 (2003) and re*ferences cited therein.*

Krampera M, et al (2002). Bone marrow mesenchymal stem cells inhibit the response of naive and memory antigen-specific T cells to their cognate peptide.

Lanzendorf SE, Boyd CA, Wright DL, Muasher S, Oehninger S. Use of human gametes obtained from anonymous donors for the production of human embryonic stem cell lines. Fertility and Sterility 2001; 76: 132-137.

Lebkowski JS, Gold J, Xu C, Funk W, Chiu CP, Carpenter MK. Human embryonic stem cells: culture, differentiation, and genetic modification for regenerative medicine applications. Cancer J 2001; 7: S83-93.

Levkau B, Koyama H, Raines EW, Clurman BE, Herren B, Orth K, Roberts JM, Ross R. Cleavage of p21cip1/waf1 and p27 kip1 mediates apoptosis in endothelial cells through activation of cdk2: role of a caspase cascade.

Li, Y., Jenkins, C.W., Nichols, M.A. and Xiong, Y. (1994) Cell cycle expression and p53 regulation of the cyclin-dependent kinase inhibitor p21. Oncogene, 9, 2261-2268 **24.** Steinman, R.A., Hoffman, B., Iro, A., Guillouf, C., Liebermann, D.A. and El-Houseini, M.E. (1994) Induction of p21 (WAF1/CIP1) during differentiation.

Liechty KW, et al (2000). Human mesenchymal stem cells engraft and demonstrate site-specific differentiation after in utero transplantation in sheep.

Lodie, T. A., et *al*., "*System*atic analysis of reportedly distinct populations of multipotent bone marrow-derived stem cells reveals a lack of distinction," Tissue Engineering 8: 739-751 (2002).

Ma Y, Ramezani A, Lewis R, Hawley R, Thomson JA. High-level sustained transgene expression in human embryonic stem cells using lentiviral vectors. Stem Cells 2003; 21: 111-117.

MacLellan WR and Schneider MD. Genetic dissection of cardiac growth control pathways Annu. Rev. Physiol. 2000. 62:289-320.

Mahon NG, O'Roke C, Codd MB, et al. Hospital mortality of acute myocardial infarction in the thrombolytic era. Heart 1999;81:478-82.

Makino S, Fukuda K, Miyoshi S, et al. Cardiomyocytes can be generated from marrow stromal cells in vitro. J Clin Invest 1999, 103:697-705.

Matzinger P (1994) Tolerance, danger, and the extended family. Annu Rev Immunol 12:991-1045.

Mellor AL, Munn DH (2001) Extinguishing maternal immune responses during pregnancy: implications for immunosuppression. Semin Immunol 13:213-218.

Moore FL, Jaruzelska J, Fox MS, Urano J, Firpo MT, Turek PJ, Dorfman DM,. Pera RA. Human Pumilio-2 is expressed in embryonic stem cells and germ cells and interacts with DAZ (Deleted in AZoospermia) and DAZ-like proteins. Proceedings of the National Academy of Sciences of the United States of America 2003; 100: 538- 543.

Nidorf SM, Siu SC, Galambos G, Weyman AE, Picard MH. Benefit of late coronary reperfusion on ventricular morphology and function after myocardial infarction. J Am Coll Cardiol 1992;20:307-313.

Pfeffer JM, Pfeffer MA, Fletcher PJ, Braunwald E. Progressive ventricular remodeling in rat with myocardial infarction. Am J Physiol 1991; 260:H1406-14.

Pfeifer A, Ikawa M, Dayn Y, Verma I. Transgenesis by lentiviral vectors: Lack of gene silencing in mammalian embryonic stem cells and preimplantation embryos. PNAS 2002; 99: 2140-2145.

Pittenger, *M. F. et al., "Multil*ineage potential of adult human mesenchymal stem cells," Science 284: 143-147 (1999); Pittenger, M., et al., "*Adult me*senchymal stem cells: Potential for muscle and tendon regeneration and use in gene therapy," Journal of Musculoskeletal and Neuronal Interactions *2: 309-320 (2002).*

Reubinoff BE, Itsykson P, Turetsky T, Pera MF, Reinhartz E, Itzik A, Ben- Hur T. Neural progenitors from human embryonic stem cells. Nat Biotechnol 2001; 19: 1134-1140.

Reubinoff BE, Pera MF, Vajta G, Trounson AO. Effective cryopreservation of human embryonic stem cells by the open pulled straw vitrification method. Hum Reprod 2001; 16: 2187-2194.

Soonpaa MH, Field LJ. Assessment of cardiomyocyte DNA synthesis in normal and injured adult mouse hearts. *Am J Physiol* **272**, H220-6 (1997).

Steinberg, D., "Can Moral Worthiness Be Seen Using a Microscope?" *American Journal of Bioethics* 2(1): 49 (2002).

6

Code of Conduct on Biotechnology

Code of Conduct on Biotechnology as it relates to Genetic Resources for Food and Agriculture

In 1991, the CGRFA requested the preparation of a draft Code of Conduct on Biotechnology, with the aim of maximizing the positive effects, and minimizing the possible negative effects, of biotechnology. The draft Code of Conduct on Biotechnology, was drawn up following a survey of over 400 international experts from the scientific community and members of the civil society. It contained five modules:

(a) biosafety and other environmental concerns;
(b) intellectual property rights and farmers rights;
(c) appropriate biotechnology for developing countries;
(d) minimizing the possible negative effects of biotechnology; and
(e) monitoring.

In 1993, noting that the Convention on Biological Diversity was considering the development of a Biosafety Protocol, the CGRFA recommended that the biosafety component of the draft Code of Conduct for Biotechnology be forwarded to it , as an input to this work. Meanwhile, the CGRFA postponed further work on the draft Code of Conduct till the completion of the negotiations for the revision of the International Undertaking on Plant Genetic Resources, which concluded in 2001. The CGRFA was kept informed by the Secretariat on the recent advancements in the field of plant and animal biotechnology. Work on the Code began again with a report on *policy issues, gaps and duplications* to the 2004 Tenth Session of the Commisison. The next session in 2006 will again discuss the draft Code of Conduct.

Bibliography

Anderlik, M. R. & Rothstein, M. A. Privacy and confidentiality of genetic information: What rules for the new science? *Annu. Rev. Genom. Hum. Genet.* 2, 401–433 (2001).

Anversa P, and Nadal-Ginard B. Myocyte renewal and ventricular remodelling Nature 415, 240—243, 2002.

Asahara, T. et al. Isolation of putative progenitor cells for endothelial angiogenesis. Science 1997; 275: 964-967.

Atkinson, M. A. and Leiter, E. H., "The NOD mouse model of type 1 diabetes: As good as it gets?" Nature Medicine 5: 601-604 (1999).

Auchincloss H Jr (2001) In search of the elusive holy grail: the mechanisms and prospects for achieving clinical transplantation tolerance. Am J Transplant 1: 6-12.

Beddington RS, Robertson EJ. An assessment of the developmental potential of embryonic stem cells in the midgestation mouse embryo. Development 1989; 105: 733-737.

Beltrami AP, et al. Evidence that human cardiac myocytes divide after myocardial infarction. *N Engl J Med* 344, 1750-7 (2001).

Berkowitz, P., "The Meaning of Federal Funding, " a paper commissioned by the President's Council on Bioethics, and included as Appendix F of this report. See also Khushf, G., and Best, R., "Stem Cells and the Man on the Moon: Should We Go There from Here?" *American Journal of Bioethics* 2(1): 38 (2002).

Bortvin, A., Eggan, K., Skaletsky, H., Akutsu, H., Berry, D. L., Yanagimachi, R., Page, D. C., and Jaenisch, R.

(2003). Incomplete reactivation of Oct4-related genes in mouse embryos cloned from somatic nuclei. Development *130*, 1673-1680.

Brent L (1997) Fetally and neonatally induced immunologic tolerance. In: Brent L (ed) A history of transplantation immunology. Academic, San Diego.

Brownstein, R., "Bush Won't Budge on Stem Cell Position, Health Secretary Says, " *Los Angeles Times*, August 13, 2001, p. A9.

Bush, G. W., "Stem Cell Science and the Preservation of Life, " *The New York Times*, August 12, 2001, p. D13.

BZ Atkins, MT Hueman, JM Meuchel, MJ Cottman, KA Hutcheson, DA Taylor: Myogenic cell transplantation improves in vivo regional performance in infarcted rabbit myocardium. J Heart Lung Transpl 1999, 18: 1173-1180.

Carpenter, Janet and Leonard Gianessi, "Herbicide Tolerant Soybeans: Why Growers Are Adopting Roundup Ready Varieties, " *AgBioForum*, vol. 2, no. 2, 1999. Available at www. agbioforum. missouri. edu.

CE Murry, RW Wiseman, SM Schwartz, SD Hauschka: Skeletal myoblast transplantation for repair of myocardial necrosis. J Clin Invest 1996, 98: 2512-2523.

Chadwick K, Wang L, Li L, Menendez P, Murdoch B, Rouleau A, Bhatia M.

Chen, Y., *et al., "Embryonic* stem cells generated by nuclear transfer of human somatic nuclei into rabbit oocytes, " Cell Research 13: 251-264 (2003).

Cheng L, Hammond H, Ye Z, Zhan X, Dravid G. Human adult marrow cells support prolonged expansion of human embryonic stem cells in culture. Stem Cells 2003; 21: 131-142.

Choi K, Kennedy M, Kazarov A, Papadimitriou, Keller G. A common precursor for hematopoietic and endothelial cells. Development 1998; 125: 725-732.

Clark, J., "Squandering Our Technological Future, " *The New York Times*, August 30, 2001, p. A19. See also the presentation of Thomas Okarma, President and CEO of Geron Corporation, before the Council on September 4, 2003. The full transcript of Okarma's presentation may be found on the Council's website at www. bioethics. gov.

Coalition, and Ethics, A. f. R. (2003). Do no harm—Reality check: proof of "therapeutic cloning"? wwwstemcellresearchorg/pr/pr_2003-03-10htm.

Collins, Francis S. "Testimony before the Presidents Council on Bioethics. " (December 13, 2002). Available at: http: //www. bioethics. gov.

Colucci WS. Molecular and cellular mechanisms of myocardial failure. Am J Cardiol 1997; 80: 15L-25L.

Council's July 2002 report *Human Cloning and Human Dignity: An Ethical Inquiry,* Chapters 6 and 8, available on the Council's website at www. bioethics. gov.

Daley, G. Q., et al., "Realistic Prospects for Stem Cell Therapeutics, " Hematology American Society for Hematology Education *Program: 39*8-418 (2003).

Daley, G., "Cloning and Stem Cells—Handicapping the Political and Scientific Debates, " *New England Journal of Medicine* 349(3): 211-212 (2003).

Druker, B. J. Imatinib alone and in combination for chronic myeloid leukemia. *Semin. Hematol.* 40, 50–8 (2003).

Drukker M, Katz G, Urbach A, Schuldiner M, Markel G, Itskovitz-Eldor J, Reubinoff B, Mandelboim O, Benvenisty N. Characterization of the expression of MHC proteins in human embryonic stem cells. Proc Natl Acad Sci U S A 2002; 99: 9864-9869.

Drukker, M., et al., *"Charac*terization of the expression of MHC proteins in human embryonic stem cells, " Proceedings of the National Academy of Sciences of the *United States of America 99: 9864-9869 (2002).*

Edwards, R. G. "IVF and the History of Stem Cells. " Nature, vol. 413 (September 27, 2001): 349-51.

Edwards, Steptoe, and Purdy. "Fertilization and Cloning In Vitro of Preovulation Human Oocytes. " Nature 277 (1970): 1307-1309.

Eggan, K., Akutsu, H., Loring, J., Jackson-Grusby, L., Klemm, M., Rideout, W.

Eggan, K., Rode, A., Jentsch, I., Samuel, C., Hennek, T., Tintrup, H., Zevnik, B., Erwin, J., Loring, J., Jackson-Grusby, L., *et al.* (2002). Male and female mice derived from the same embryonic stem cell clone by tetraploid embryo complementation.

Eiges R, Schuldiner M, Drukker M, Yanuka O, Itskovitz-Eldor J, Benvenisty N. Establishment of human embryonic stem cell-transfected clones carrying a marker for undifferentiated cells. Current Biology 2001; 11: 514-518.

Elefanty AG, Robb L, Birner R, Begley CG. Hematopoietic-specific genes are not induced during in vitro differentiation of scl-null embryonic stem cells.

Faraci, M., et *al., "Severe ne*urologic complications after hematopoietic stem cell transplantation in children, " Neurology 59: 1895-1904 (2002).

Farley, Margaret A. "Roman Catholic Views on Research Involving Human Embryonic Stem Cells. " The Human Embryonic Stem Cell Debate: Science, Ethics, and Public Policy. Cambridge, MA: The MIT Press, 2001. 113-18.

Feeder-free growth of undifferentiated human embryonic stem cells. Nat Biotechnol 2001; 19: 971-974.

Ferguson-Smith, A. C., and Surani, M. A. (2001). Imprinting and the epigenetic asymmetry between parental genomes. Science *293*, 1086-1089.

Finger S, Heavens RP, Sirinathsinghji DJ, Kuehn MR, Dunnett SB.. Behavioral and neurochemical evaluation of a transgenic mouse model of Lesch-Nyhan syndrome. Journal of Neurological Science 1988; 86: 203-213.

FitzPatrick, William. "Surplus Embryos, Nonreproductive Cloning, and the Intend/ Foresee Distinction. " Hastings Center Report (May-June 2003): 29-36.

Folkman, J. Therapeutic angiogenesis in ischemic limbs. Circulation 1998; 97: 108- 110.

Frazier OH, Rose EA, Macmanus Q, et al. Multicenter clinical evaluation of the Heartmate 1000 IP left ventricular assist device. Ann Thorac Surg 1992; 102: 578-587.

Fukuyama, Francis. Our Posthuman Future. New York: Farrar, Straus and Giroux, 2002.

Geron Ethics Advisory Board. "Research with Human Embryonic Stem Cells: Ethical Considerations. " Hastings Center Report, vol. 29, no. 2 (March-April 1999).

Gianessi, Leonard P. and Janet E. Carpenter, "Agricultural Biotechnology: Insect Control Benefits, " National Center for Food and Agricultural Policy, 1999. Available on Biotechnology Industry Organization at www. bio. org.

Giles, J., and Knight, J. (2003). Dolly's death leaves researchers woolly on clone ageing issue. Nature *421*, 776.

Gilliam, Lynn. "Arguing by Analogy in the Fetal Tissue Debate. " Bioethics, vol. 11, no. 5 (1997): 397-412.

Glannon, Walter. "Extending the Human Life Span. " Journal of Medicine and Philosophy, vol. 27, no. 3 (2002): 339-54.

Goldstein RS, Drukker M, Reubinoff BE, Benvenisty N. Integration and differentiation of human embryonic stem cells transplanted to the chick embryo.

Golub, T. R. Genomic approaches to the pathogenesis of hematologic malignancy. *Curr. Opin. Hematol.* 8, 252–261 (2001).

Green, R., "Determining Moral Status, " *American Journal of Bioethics* 2(1): 28-29 (2002).

Green, Ronald M. "Benefiting from 'Evil': An Incipient Moral Problem in Human Stem Cell Research. " Bioethics, vol. 16, no. 6 (2002): 544-56.

Gronthos, S., *et al., "Molecul*ar and cellular characterization of highly purified stromal stem cells derived from bone marrow, " Journal of Cell Science 116: 1827-1835 (2003).

Guttmacher and Mark S. Guyer are at the National Human Genome Research Institute, National Institutes of Health, Bethesda, Maryland 20892, USA.

Guttmacher, A. E. & Collins, F. S. Genomic medicine — A primer. *N. Engl. J. Med.* 347, 1512–1520 (2002).

Halevy, O., Novitch, B. G., Spicer, D. B., Skapek, S. X., Rhee, J., Hannon, G. J., Beach, D. and Lassar, A. B. (1995) Correlation of terminal cell cycle arrest of skeletal muscle with induction of p21 by MyoD. Science, 267, 1018- 1021.

Hall, Stephen S. Merchants of Immortality: Chasing the Dream of Human Life Extension. New York: Houghton Mifflin Company, 2003.

Hannon, G. J. RNA interference. *Nature* 418, 244–251 (2002).

Harris, J., "Stem Cells, Sex & Procreation, " a paper presented to the American Philosophical Society, March 29, 2003.

He JQ, Ma Y, Lee Y, Thomson JA, Kamp TJ. Human Embryonic Stem Cells Develop Into Multiple Types of Cardiac Myocytes. Action Potential Characterization.

Hematopoietic colony-forming cells derived from human embryonic stem cells. Proc Natl Acad Sci U S A 2001; 98: 10716-10721.

Henderson JK, Draper JS, Baillie HS, Fishel S, Thomson JA, Moore H, Andrews PW. Preimplantation human embryos and embryonic stem cells show comparable expression of stage-specific embryonic antigens. Stem Cells 2002; 20: 329-337.

Hescheler J, Fleischmann BK, Wartenberg M, Bloch W, Kolossov E, Ji G, Addicks K, Sauer H (1999) Establishment of ionic channels and signalling cascades in the embryonic stem cell-derived primitive endoderm and cardiovascular system. Cells Tissues Organs 165: 153-164.

Heymans S, Luutun A, Nuyens D, et al. Inhibition of plasminogen activators or matrix metalloproteinases prevents cardiac rupture but impairs therapeutic angiogenesis and causes cardiac failure. Nat Med 1999; 5: 1135- 1142.

Hilbert, D. Mathematical problems. *Bull. Am. Math. Soc.* 8, 437–479 (1902).

Hill MF, Singal PK. Right and left myocardial antioxidant responses during heart failure subsequent to myocardial infarction. Circulation 1997 96: 2414-20.

Hirschhorn, J. N., Lohmueller, K., Byrne, E. & Hirschhorn, K. A comprehensive review of genetic association studies. *Genet.*

Hochman JS and Choo H. Limitation of myocardial infarct expansion by reperfusion independent of myocardial salvage. Circulation 1987; 75: 299- 306.

Hodgetts SI, Beilharz MW, Scalzo T, Grounds MD (2000) Why do cultured transplanted myoblasts die in vivo? DNA quantification shows enhanced survival of donor male myoblasts in host mice depleted of CD4+ and CD8+ or NK1. 1+ cells. Cell Transplant 9: 489-502.

Hoffman, D. I., et al., "Cryopreserved Embryos in the United States and Their Availability for Research, " *Fertility and Sterility* 79: 1063-1069 (2003).

Hoffman, David I. et al. "Cryopreserved Embryos in the United States and Their Availability for Research. " Fertility and Sterility, vol. 79, no. 5 (May 2003): 1063- 69.

Hogan, P., et al., "Economic Costs of Diabetes in the US in 2002, " Diabetes Care 26: 917-932 (2003).

Hognes JR. In The Artificial Heart: Prototypes, Policies and Patients. Washingtion, DC: National Academy Press; 1991: 1-312.

Holland, Suzanne. "Contested Commodities at Both Ends of Life: Buying and Selling Gametes, Embryos, and Body Tissues, " Kennedy Institute of Ethics Journal, vol. 11, no. 3 (2001): 266.

Holt, R. A. *et al.* The genome sequence of the malaria mosquito *Anopheles gambiae*. *Science* 298, 129–149 (2002).

Horii, T., Nagao, Y., Tokunaga, T., and Imai, H. 20003. Serum-free culture of murine primordial germ cells and embryonic germ cells. Theriogen. 59, 1257-1264.

Horii, T., Nagao, Y., Tokunaga, T., and Imai, H. 20003. Serum-free culture of murine primordial germ cells and embryonic germ cells. Theriogen. 59, 1257-1264.

Hubner, K., Fuhrmann, G., Christenson, L. K., Kehler, J., Reinbold, R., De La Fuente, R., Wood, J., Strauss, I. J., Boiani, M., and Scholer, H. R. (2003). Derivation of Oocytes from Mouse Embryonic Stem Cells. Science.

Human bone marrow stem cells exhibit neural phenotypes and ameliorate neurological deficits after grafting into the ischemic brain of rats. Exp Neurol.

Humpherys, D., Eggan, K., Akutsu, H., Friedman, A., Hochedlinger, K., Yanagimachi, R., Lander, E., Golub, T. R., and Jaenisch, R. (2002). Abnormal gene expression in cloned mice derived from ES cell and cumulus cell nuclei. Proc Natl Acad Sci U S A *99*, 12889-12894.

Jackson KA, Majka SM, Wang H, Pocius J, Hartley CJ, Majesky MW, Entman ML, Michael LH, Hirschi KK, Goodell MA (2001) Regeneration of ischemic cardiac muscle and vascular endothelium by adult stem cells. J Clin Invest 107: 1395-1402.

Jackson, D. A., Symons, R. H. & Berg, P. Biochemical method for inserting new genetic information into DNA of Simian Virus 40: circular SV40 DNA molecules containing lambda phage genes and the galactose operon of *Escherichia coli. Proc. Natl Acad. Sci.*

Jaenisch, R., and Wilmut, I. (2001). Developmental biology. Don't clone humans! Science *291*, 2552.

Jaffredo T, Gautier R, Eichmann A, Dieterlen-Lievre F. Intraaortic hemopoietic cells are derived from endothelial cells during ontogeny. Development 1998; 125: 4575-4583.

James, Clive, "Global Review of Commercialized Transgenic Crops: 1998, " *ISAAA*, No. 8, 1998.

Kaczorowski, D. J., Patterson, E. S., Jastromb, W. E., Shamblott, M. J. 2002. Review Article. Glucose-responsive insulin-producing cells from stem cells. Diabetes Metab.

Kaczorowski, D. J., Patterson, E. S., Jastromb, W. E., Shamblott, M. J. 2002. Review Article. Glucose-responsive insulin-producing cells from stem cells. Diabetes Metab.

Kahn, J., "Missing the mark on stem cells, " *Bioethics Examiner* 5: 3, Fall 2001, p. 1.

Kajstura J, Leri A, Finato N, di Loreto N, Beltramo CA, Anversa P. Myocyte proliferation in end-stage cardiac failure in humans. Proc Natl Acad Sci USA 95, 8801-8805 (1998).

Kalka, C. et al. Transplantation of ex vivo expanded endothelial progenitor cells for therapeutic neovascularization. Proc. Natl. Acad. Sci. USA 2000; 97: 3422-3427.

Kalkman EAJ, Bilgin YM, van Haren P, van Suylen R-J, Saxena PR, Schoemaker RG. Determinants of coronary reserve in rats subjected to coronary artery ligation or aortic banding. Cardiovasc Res 1996.

Kass, L., "The Meaning of Life—In the Laboratory, " *The Public Interest*, Winter 2002. Also see the Council's

July 2002 report *Human Cloning and Human Dignity: An Ethical Inquiry,* Chapter 6.

Kass, Leon R. "Ageless Bodies, Happy Souls: Biotechnology and the Pursuit of Perfection. " The New Atlantis, no. 1 (spring 2003).

Kaveny, M. Cathleen. "Appropriation of Evil: Cooperation's Mirror Image, " Theological Studies, vol. 61, no. 2 (June 2000): 280-313.

Kawano H, Do YS, Kawano Y, Starnes V, Barr M, Law RE, Hsueh WA. Angiotensin II has multiple profibrotic effects in human cardiac fibroblasts. Circulation 2000; 101: 1130-1137.

Kehat I, Gepstein A, Spira A, Itskovitz-Eldor J, Gepstein L. High-resolution electrophysiological assessment of human embryonic stem cell-derived cardiomyocytes: a novel in vitro model for the study of conduction. Circ Res 2002; 91: 659-661.

Kehat I, Kenyagin-Karsenti D, Snir M, Segev H, Amit M, Gepstein A, Livne E, Binah O, Itskovitz-Eldor J, Gepstein L. Human embryonic stem cells can differentiate into myocytes with structural and functional properties of cardiomyocytes. J Clin Invest 2001; 108: 407-414.

Kehat I, Kenyagin-Karsenti D, Snir M, Segev H, Amit M, Gepstein A, Livne E, Binah O, Itskovitz-Eldor J, Gepstein L. Human embryonic stem cells can differentiate into myocytes with structural and functional properties of cardiomyocytes. J Clin Invest. 2001; 108: 407-14.

Kennedy M, Firpo M, Choi K, Wall C, Robertson S, Kabrun N, Keller G. A common precursor for primitive erythropoiesis and definitive haematopoiesis.

Kennedy, D., "Stem Cells: Still Here, Still Waiting, " *Science* 300: 865 (2003).

Kerr, D. A., et al., "*Hum*an Embryonic Germ Cell Derivatives Facilitate Motor Recovery of Rats with Diffuse Motor Neuron Injury, " The Journal of Neuroscience 23: 5131-5140 (2003).

Kerr, D. A., Lladó, J., Shamblott, M. J., Maragakis, N. J., Irani, D. N., Crawford, T. O, Krishnan, C., Dike, S., Gearhart, J. D., and Rothstein, J. D. 2003. Human embryonic germ cell derivatives facilitate motor recovery of rats with diffuse motor neuron injury. J. Neurosci. 23, 5131-5140.

Kerr, D. A., Lladó, J., Shamblott, M. J., Maragakis, N. J., Irani, D. N., Crawford, T. O, Krishnan, C., Dike, S., Gearhart, J. D., and Rothstein, J. D. 2003. Human embryonic germ cell derivatives facilitate motor recovery of rats with diffuse motor neuron injury. J. Neurosci. 23, 5131-5140.

Khushf, G., and Best, R., *op. cit.,* p. 38.

Kimura, T., Suzuki, A., Fujita, Y., Yomogida, K., Lomeli, H., Asada, N., Ikeuchi, M., Nagy, A., Mak, T. W., and Nakano, T. 2003. Development and Disease. Conditional loss of PTEN leads to testicular teratoma and enhances embryonic germ cell production. Development 130, 1691-1700.

Kimura, T., Suzuki, A., Fujita, Y., Yomogida, K., Lomeli, H., Asada, N., Ikeuchi, M., Nagy, A., Mak, T. W., and Nakano, T. 2003. Development and Disease. Conditional loss of PTEN leads to testicular teratoma and enhances embryonic germ cell production. Development 130, 1691-1700.

Klug MG, Soonpaa MH, Koh GY, Field LJ (1996) Genetically selected cardiomyocytes from differentiating embryonic stem cells form stable intracardiac grafts. J Clin Invest 98: 216-224.

Knowles, Lori P. "Property, Progeny, and Patents. " Hastings Center Report, vol 29.

Koc, O. N., et al., "*Rapid he*matopoietic recovery after coinfusion of autologousblood stem cells and culture-expanded marrow mesenchymal cells in advanced breast cancer patients receiving high-dose chemotherapy, " Journal of Clinical Oncology 18: 307-316 (2000).

Kocher A, Schuster M, Szabolcs M, Itescu S. Cardiomyocyte regeneration after neovascularization of ischemic myocardium by human bone marrowderived angioblasts. Nature Medicine 2002, in press.

Kocher AA, Schuster MD, Szabolcs MJ, Takuma S, Burkhoff D, Wang J, Homma S, Edwards NM, Itescu S. Neovascularization of ischemic myocardium by human bone-marrow-derived angioblasts prevents cardiomyocyte apoptosis, reduces remodeling and improves cardiac function. Nature Med.

Kondo, M*., et al*., "*Biology of H*ematopoietic Stem Cells and Progenitors: Implications for Clinical Application, " Annual Review of Immunology 21: 759-806 (2003) and re*ferences cited therein.*

Krampera M, et al (2002). Bone marrow mesenchymal stem cells inhibit the response of naive and memory antigen-specific T cells to their cognate peptide.

Lauritzen, P., "Report on the Ethics of Stem cell research, " a commissioned paper prepared at the request of the Council and included in Appendix G of this report.

Lauritzen, Paul. "Neither Person nor Property: Embryo Research and the Status of the Early Embryo, " *America* (March 26, 2001).

Lebacqz, Karen. "Stem Cells and Justice, " Dialog, vol. 41, no. 3 (fall 2002).

Lebkowski JS, Gold J, Xu C, Funk W, Chiu CP, Carpenter MK. Human embryonic stem cells: culture, differentiation, and genetic modification for regenerative medicine applications. Cancer J 2001; 7: S83-93.

Levenberg S, Golub JS, Amit M, Itskovitz-Eldor J, Langer R. Endothelial cells derived from human embryonic stem cells. Proc Natl Acad Sci U S A 2002; 99: 4391-4396.

Levkau B, Koyama H, Raines EW, Clurman BE, Herren B, Orth K, Roberts JM, Ross R. Cleavage of p21cip1/waf1 and p27 kip1 mediates apoptosis in endothelial cells through activation of cdk2: role of a caspase cascade.

Lewis, C. S. The Abolition of Man. New York: The Macmillan Company, 1947.

Liechty KW, et al (2000). Human mesenchymal stem cells engraft and demonstrate site-specific differentiation after in utero transplantation in sheep.

Liechty KW, MacKenzie TC, Shaaban AF, et al. Human mesenchymal stem cells engraft and demonstrate site-specific differentiation after in utero transplantation in sheep. Nat Med 2000, 6: 1282-6.

Liker, M., et al., "Huma*n neura*l stem cell transplantation in the MPTP-lesioned mouse, " Brain Research 971: 168-177 (2003).

Lin Y, Weisdorf DJ, Solovey A, Hebbel RP. Origins of circulating endothelial cells and endothelial outgrowth from blood. J Clin Invest 2000; 105: 71-77.

Lodie, T. A., et a*l., "System*atic analysis of reportedly distinct populations of multipotent bone marrow-derived stem cells reveals a lack of distinction, " Tissue Engineering 8: 739-751 (2002).

Los Angeles, CA. : University of California Press, 2003; also see Daniel Callahan's presentation before the Council on July 24, 2003, available on the Council's website at www. bioethics. gov.

Lysaght, M. J., and Hazlehurst, A. L., "Private Sector Development of Stem Cell Technology and Therapeutic Cloning, " *Tissue Engineering* 9(3): 555-561 (2003).

Ma Y, Ramezani A, Lewis R, Hawley R, Thomson JA. High-level sustained transgene expression in human embryonic stem cells using lentiviral vectors. Stem Cells 2003; 21: 111-117.

Macklin, Ph. D., Ruth. "Ethics, Politics, and Human Embryo Stem Cell Research. " Women's Health Issues, vol. 10, no. 3 (May/June 2000): 111-15.

MacLellan WR and Schneider MD. Genetic dissection of cardiac growth control pathways Annu. Rev. Physiol. 2000. 62: 289-320.

Magnus, David, Arthur Caplan, and Glenn McGee, ed. In Who Owns Life? Amherst, New York: Prometheus Books, 2002.

Mahon NG, O'Roke C, Codd MB, et al. Hospital mortality of acute myocardial infarction in the thrombolytic era. Heart 1999; 81: 478-82.

Mahowald, M. and Mahowald A., "Embryonic Stem Cell Retrieval and a Possible Ethical Bypass, " *American Journal of Bioethics* 2(1): 42-43 (2002).

Makino S, Fukuda K, Miyoshi S, et al. Cardiomyocytes can be generated from marrow stromal cells in vitro. J Clin Invest 1999, 103: 697-705.

Maxam, A. M. & Gilbert, W. A new method for sequencing DNA.

Mayer, W., Niveleau, A., Walter, J., Fundele, R., and Haaf, T. (2000). Demethylation of the zygotic paternal genome. Nature *403*, 501-502.

McCormick, R. A. "Who or What is the Preembryo?" In Corrective Vision: Explorations in Moral Theology. Kansas City: Sheed & Ward, 1994.

McEwan PE, Gray GA, Sherry L, Webb DJ, Kenyon CJ. Differential effects of angiotensin II on cardiac cell proliferation and intramyocardial perivascular fibrosis in vivo. Circulation 1998; 98: 2765-2773.

McGee, Glenn, and Elizabeth Banger. "Ethical Issues in the Patenting and Control of Stem Cell Research. " In

Who Owns Life? Ed. David Magnus, Arthur Caplan, and Glenn McGee. Amherst, New York: Prometheus Books, 2002. 243-264.

McKenny, Gerald P. "Enhancements and the Ethical Significance of Vulnerability. " In Enhancing Human Traits: Ethical and Social Implications. Washington, D. C. : Georgetown University Press, 1998. 222-37.

McKibben, Bill. Enough: Staying Human in an Engineered Age. New York: Henry Holt and Company, 2003.

McKiernan SH, Bavister BD. Different lots of bovine serum albumin inhibit or stimulate in vitro development of hamster embryos. In Vitro Cell Dev Biol 1992; 28A: 154-156.

Meilaender, Gilbert. "Terra es animata: On Having a Life. " Hastings Center Report, vol. 23, no. 4 (July-August 1993): 25-32.

Mellor AL, Munn DH (2001) Extinguishing maternal immune responses during pregnancy: implications for immunosuppression. Semin Immunol 13: 213-218.

Menasche P, Hagege AA, Scorsin M, Pouzet B, Desnos M, Duboc D, Schwartz K, Vilquin J-T, Marolleau J-P (2001) Myoblast transplantation for heart failure.

Mendel, G. Versuche über Pflanzen-Hybriden. *Verhandlungen des naturforschenden Vereines, Abhandlungen, Brünn* 4, 3–47 (1866).

Mesenchymal stem cells inhibit and stimulate mixed lymphocyte cultures and mitogenic responses independently of the major histocompatibility complex. Scand J Immunol. 57(1): 11-20.

Meyer, Michael J., and Lawrence J. Nelson. "Respecting What We Destroy: Reflections on Human Embryo Research. " Hastings Center Report, vol. 31, no.

Midgley, Mary. "Biotechnology and Monstrosity: Why We Should Pay Attention to the 'Yuk Factor. '" Hastings Center Report, vol. 30, no. 5 (2000): 7-15.

Migeon, B. R., Chowdry, A. K., Dunston, J. A., and McIntosh, I. 2001. Identification of *TSIX*, encoding an RNA antisense to human *XIST*, reveals differences from its murine counterpart: Implications for X inactivation. Am. J. Human Genet. 69, 951-960.

Milki, A. A. et al., "Incidence of monozygotic twinning with blastocyst transfer compared to cleavage-stage transfer" *Fertility and Sterility* 79(3): 503-506 (2003).

Multipotent progenitor cells can be isolated from post-natal murine bone marrow, muscle and brain. Exp Hematol. 2002; 30(8): 896-904 5. Reyes M, Dudek A, Jahagirdar B, Koodie K, Marker PH, Verfaillie CM.

Mummery C, Ward D, van den Brink CE, Bird SD, Doevendans PA, Opthof T, Brutel de la Riviere A, Tertoolen L, van der Heyden M, Pera M. Cardiomyocyte differentiation of mouse and human embryonic stem cells. J Anat 2002; 200: 233- 242.

Norwitz, E. R., et al., "Implantation and the survival of early pregnancy, " *New England Journal of Medicine* 345(19): 1400-1408 (2001)

Nussbaum, Martha C. "Compassion and Terror. " Daedalus, vol. 128, no. 4 (winter 2003): 10-26.

Odorico JS, Kaufman DS, Thomson JA. Multilineage differentiation from human embryonic stem cell lines. Stem Cells 2001; 19: 193-204.

Ogawa M, Kizumoto M, Nishikawa S, Fujimoto T, Kodama H, Nishikawa SI. Blood 1999; 93: 1168-1177.

Ogonuki, N., Inoue, K., Yamamoto, Y., Noguchi, Y., Tanemura, K., Suzuki, O., Nakayama, H., Doi, K., Ohtomo, Y., Satoh, M., *et al.* (2002). Early death of mice cloned from somatic cells. Nat Genet *30*, 253-254.

Okarma, T., Presentation at the September 4, 2003, meeting of the President's Council on Bioethics, Washington, D. C., available at www. bioethics. gov.

Onyango, P., Jiang, S., Uejima, H., Shamblott, M. J., Gearhart,

J. D., Cui, H., and Feinberg, A. P. 2002. Monoallelic expression and methylation of imprinted genes in human and mouse embryonic germ cell lineages. Proc. Natl. Acad. Sci. USA 99, 10599-10604.

Orlic D, Kajstura J, Chimenti S, et al. Bone marrow cells regenerate infarcted myocardium. Nature 2001, 410: 701-705.

Orr, R., "The Moral Status of the Embryonal Stem Cell: Inherent or Imputed?" *American Journal of Bioethics* 2(1): 57-59 (2002).

Orr, R., *op. cit.,* pp. 57-58; "A review of the National Institute of Health's Guidelines for Research Using Human Pluripotent Stem Cells, " *Issues in Law and Medicine* 3(17): 293.

Outka, G., "The Ethics of Stem cell research, " a paper presented to the Council for its April 2002 meeting, and available on the Council website at www. bioethics. gov. A slightly revised version appeared in the *Kennedy Institute of Ethics Journal*, 12: 2, pp. 175-213.

Outka, Gene. "The Ethics of Stem Cell Research. " Kennedy Institute of Ethics Journal vol. 12, no. 2 (2002): 175-213.

Pagano, S. F., *et al.*, "*Is*olation and characterization of neural stem cells from the adult human olfactory bulb, " Stem Cells 18: 295-300 (2000).

Park, T. S., Hong, Y. H., Kwon, S. C., Lim, J. M., and Han, J. Y. 2003. Birth of germline chimeras by transfer of chicken embryonic germ (EG) cells into recipient embryos.

Pfeffer MA, Braunwald E, Moye LA, et al. Effect of captopril on mortality and morbidity in patients with left ventricular dysfunction after myocardial infarction. Results of the survival and ventricular enlargement trial. The SAVE investigators. N Engl J Med 1992; 327: 669-677.

Pfeifer A, Ikawa M, Dayn Y, Verma I. Transgenesis by lentiviral vectors: Lack of gene silencing in mammalian embryonic stem cells and preimplantation embryos. PNAS 2002; 99: 2140-2145.

Pitt B, Segal R, Martinez FA, et al. Randomised trial of losartan versus captopril in patients over 65 with heart failure (Evaluation of Losartan in the Elderly Study, ELITE). Lancet 1997; 349: 747-752.

Pittenger MF, Mackay AM, Beck SC, et al. Multilineage potential of adult human mesenchymal stem cells. Science 1999, 284: 143-147.

Pontifical Academy for Life. "On the Production and the Scientific and Therapeutic Use of Human Embryonic Stem Cells. " Vatican (August 25, 2000): available at Http: //www. petersnet. net/browse/ 3021. htm.

Potential approaches for myocardial regeneration. Ann N Y Acad Sci 1995; 752: 446- 454.

Pouzet B, Ghostine S, Vilquin J-T, Garcin I, Scorsin M, Hagege AA, Duboc D, Schwartz K, Menasche P (2001) Is skeletal myoblast transplantation clinically relevant in the era of angiotensin-converting enzyme inhibitors? Circulation 104: I223-228.

Proc. Natl. Acad. Sci U S A. 2003; In Press 8. Ling Z, Potter E, Lipton J, Carvey P. Differentiation of mesencephalic progenitor cells into dopaminergic neurons by cytokines. Exp Neurol.

Pursley, W. H., Presentation at the September 4, 2003, meeting of the President's Council on Bioethics, Washington, D. C., available at www. bioethics. gov.

R. Alta Charo, of the University of Wisconsin Law School, quoted in Coyle, M., "The Clone Zone, " *The National Law Journal*, May 15, 2002.

Rabinow, Paul. French DNA: Trouble in Purgatory. Chicago, IL: The University of Chicago Press, 1999.

Radin, Margaret Jane. Contested Commodities. Cambridge, MA: Harvard University Press, 1996.

Rafii S, Shapiro F, Rimarachin J, Nachman R, Ferris B, Weksler B, Moore AS, Asch AS. Isolation and characterization of human bone marrow microvascular endothelial cells: hematopoietic progenitor cell adhesion.

Rambhatla L, Chiu CP, Kundu P, Peng Y, Carpenter MK. Generation of hepatocyte-like cells from human embryonic stem cells. Cell Transplant 2003; 12: 1- 11.

Ravichandran LV and Puvanakrishnan R. In vivo labeling studies on the biosynthesis and degradation of collagen in experimental myocardial myocardial infarction. Biochem Intl 1991; 24: 405-414.

Reik, W., Dean, W., and Walter, J. (2001). Epigenetic reprogramming in mammalian development. Science *293*, 1089-1093.

Resnik, David. "The Commercialization of Human Stem Cells: Ethical and Policy Issues. " Health Care Analysis, vol. 10 (2002): 127-54.

Reubinoff BE, Itsykson P, Turetsky T, Pera MF, Reinhartz E, Itzik A, Ben- Hur T. Neural progenitors from human embryonic stem cells. Nat Biotechnol 2001; 19: 1134-1140.

Reubinoff BE, Pera MF, Vajta G, Trounson AO. Effective cryopreservation of human embryonic stem cells by the open pulled straw vitrification method. Hum Reprod 2001; 16: 2187-2194.

Rideout, W., et al., *"Correct*ion of a genetic defect by nuclear transplantation and combined cell and gene therapy, " Cell 109: 17-27 (2002).

Robertson, John A. "Ethics and Policy in Embryonic Stem Cell Research. " Kennedy Institute of Ethics Journal, vol. 9, no. 2 (1999): 109-36.

Roche, J. D., Patricia A. and Michael A. Grodin, M. D., F. A. A. P. "The Ethical Challenge of Stem Cell Research." Women's Health Issues, vol. 10, no. 3 (May/June 2000): 136-39.

Rohwed*el, J., et al., "Embr*yonic stem cells as an in vitro model for mutagenicity, cytotoxicity, and embryotoxicity studies: present state and future prospects," Toxicology In Vitro 15: 741-753 (2001).

Rothenberg, K. *et al.* Genetic information and the workplace: Legislative approaches and policy challenges. *Science* 275, 1755–1757 (1997).

Ryan, E. A., et *al., "Clinica*l outcomes and insulin secretion after islet transplantation with the Edmonton protocol, " Diabetes 50: 710-719 (2001).

Salter-Cid L, Flajnik MF (1995) Evolution and developmental regulation of the major histocompatibility complex. Crit Rev Immunol 15: 31-75.

Sanger, F. & Coulson, A. R. A rapid method for determining sequences in DNA by primed synthesis with DNA polymerase.

Sapienza, C. 2002. Commentary. Imprinted gene expression, transplantation medicine, and the "other" human embryonic stem cell. Proc. Natl. Acad. Sci. USA 99, 10243-10245.

Sapienza, C. 2002. Commentary. Imprinted gene expression, transplantation medicine, and the "other" human embryonic stem cell. Proc. Natl. Acad. Sci. USA 99, 10243-10245.

Sathananthan H, Pera M, Trounson A. The fine structure of human embryonic stem cells. Reprod Biomed Online 2002; 4: 56-61.

Smythe GM, Hodgetts SI, Grounds MD (2001) Problems and solutions in myoblast transfer therapy. J Cell Mol Med 5: 33-47.

Tsai FY, Keller G, Kuo FC, Weiss M, Chen J, Rosenblatt M, Alt FA, Orkin SH. An early hematopoietic defect in mice lacking the transcription factor GATA- 2. Nature 1994; 371: 221-225.

Van der Weyden, L., Adams, D. J. & Bradley, A. Tools for targeted manipulation of the mouse genome. *Physiol. Genomics* 11, 133–164 (2002).

Van Overwalle, Geertrui. Study on the Patenting of Inventions Related to Human Stem Cell Research. Luxembourg: Office for Official Publications of the European Communities, 2002. 23.

Waldby, Ph. D., Catherine, and Susan Squier, Ph. D. "Ontogeny, Ontology, and Phylogeny: Embryonic Life and Stem Cell Technologies. " Configurations (Forthcoming).

Weissman, I. L., "Stem Cells—Scientific, Medical and Political Issues, " *New England Journal of Medicine* 346(20): 1576-1579 (2002).

Wekerle T, Kurtz J, Ito H, Ronquillo JV, Dong V, Zhao G, Shaffer J, Sayegh MH, Sykes M (2000) Allogeneic bone marrow transplantation with co-stimulatory blockade induces macrochimerism and tolerance without cytoreductive host treatment. Nat Med 6: 464-469.

Wekerle T, Sykes M. Mixed chimerism as an approach for the induction of transplantation tolerance. Transplantation 1999; 68: 459-467.

White HD, Cross DB, Elliot JM, et al. Long-term prognostic importance of patency of the infarct-related coronary artery after thrombolytic therapy for myocardial infarction. Circulation 1994; 89: 61-67.

White HD, Norris RM, Brown MA, Brandt PWT, Whitlock RML, Wild CJ. Left ventricular end systolic volume as the major determinant of survival after recovery from myocardial infarction. Circulation 1987; 76: 44-51.

Wilcox, A. J., et al., "Incidence of early loss of pregnancy, " *New England Journal of Medicine* 319(4): 189-194 (1988).

Williams, Bernard. Problems of the Self: Philosophical Papers 1956-1972. Cambridge, England: Cambridge University Press, 1973.

Wilmut, I. (2001). How safe is cloning? Cloning *3*, 39-40.

Xu, R. H., et al., "BMP4 initiates human embryonic stem cell differentiation to trophoblast" *Nature Biotechnology* 20(12): 1261-1264 (2002).

Young, L. E., Sinclair, K. D., and Wilmut, I. (1998). Large offspring syndrome in cattle and sheep. Rev Reprod *3*, p155-163.

Zhang, M. Q. Computational prediction of eukaryotic proteincoding genes. *Nature Rev. Genet.* 3, 698–709 (2002).

Zhao, M., et al., "*Amelioration* of streptozotocin-induced diabetes in mice using human islet cells derived from long-term culture in vitro, " Transplantation 73: 1454-1460 (2002).

Zoloth, Laurie. "Jordon's Banks: A View from the First Years of Human Embryonic Stem Cell Research. " In The Human Embryonic Stem Cell Debate. Ed. by Suzanne Holland, Karen Lebacqz, and Laurie Zoloth. Cambridge, MA: The MIT Press, 2001. 238.

Zulewski, H., *et al., "Mul*tipotential Nestin-Positive Stem Cells Isolated From Adult Pancreatic Islets Differentiate Ex Vivo Into Pancreatic Endocrine, Exocrine and Hepatic Phenotypes, " Diabetes 50: 521-533 (2001).

Index

L

M

N

O

P

Q

R